T0180553

Advances in Intelligent Systems and Computing

Volume 453

Series editor

Janusz Kacprzyk, Polish Academy of Sciences, Warsaw, Poland
e-mail: kacprzyk@ibspan.waw.pl

About this Series

The series "Advances in Intelligent Systems and Computing" contains publications on theory, applications, and design methods of Intelligent Systems and Intelligent Computing. Virtually all disciplines such as engineering, natural sciences, computer and information science, ICT, economics, business, e-commerce, environment, healthcare, life science are covered. The list of topics spans all the areas of modern intelligent systems and computing.

The publications within "Advances in Intelligent Systems and Computing" are primarily textbooks and proceedings of important conferences, symposia and congresses. They cover significant recent developments in the field, both of a foundational and applicable character. An important characteristic feature of the series is the short publication time and world-wide distribution. This permits a rapid and broad dissemination of research results.

Advisory Board

Chairman

Nikhil R. Pal, Indian Statistical Institute, Kolkata, India
e-mail: nikhil@isical.ac.in

Members

Rafael Bello, Universidad Central "Marta Abreu" de Las Villas, Santa Clara, Cuba
e-mail: rbellop@uclv.edu.cu

Emilio S. Corchado, University of Salamanca, Salamanca, Spain
e-mail: escorchado@usal.es

Hani Hagras, University of Essex, Colchester, UK
e-mail: hani@essex.ac.uk

László T. Kóczy, Széchenyi István University, Győr, Hungary
e-mail: koczy@sze.hu

Vladik Kreinovich, University of Texas at El Paso, El Paso, USA
e-mail: vladik@utep.edu

Chin-Teng Lin, National Chiao Tung University, Hsinchu, Taiwan
e-mail: ctlin@mail.nctu.edu.tw

Jie Lu, University of Technology, Sydney, Australia
e-mail: Jie.Lu@uts.edu.au

Patricia Melin, Tijuana Institute of Technology, Tijuana, Mexico
e-mail: epmelin@hafsamx.org

Nadia Nedjah, State University of Rio de Janeiro, Rio de Janeiro, Brazil
e-mail: nadia@eng.uerj.br

Ngoc Thanh Nguyen, Wroclaw University of Technology, Wroclaw, Poland
e-mail: Ngoc-Thanh.Nguyen@pwr.edu.pl

Jun Wang, The Chinese University of Hong Kong, Shatin, Hong Kong
e-mail: jwang@mae.cuhk.edu.hk

More information about this series at http://www.springer.com/series/11156

Thanh Binh Nguyen · Tien Van Do
Hoai An Le Thi · Ngoc Thanh Nguyen
Editors

Advanced Computational Methods for Knowledge Engineering

Proceedings of the 4th International
Conference on Computer Science, Applied
Mathematics and Applications, ICCSAMA
2016, 2–3 May, 2016, Vienna, Austria

 Springer

Editors

Thanh Binh Nguyen
International Institute for Applied Systems
    Analysis (IIASA)
Laxenburg
Austria

Tien Van Do
Department of Networked Systems
    and Services
Budapest University of Technology
    and Economics
Budapest
Hungary

Hoai An Le Thi
Laboratory of Theoretical and Applied
    Computer Science (LITA), UFR MIM
University of Lorraine
Ile du Saulcy, Metz
France

Ngoc Thanh Nguyen
Institute of Informatics
Wrocław University of Technology
Wrocław
Poland

ISSN 2194-5357   ISSN 2194-5365   (electronic)
Advances in Intelligent Systems and Computing
ISBN 978-3-319-38883-0   ISBN 978-3-319-38884-7   (eBook)
DOI 10.1007/978-3-319-38884-7

Library of Congress Control Number: 2016938661

This Springer imprint is published by Springer Nature
The registered company is Springer International Publishing AG Switzerland

Preface

This volume contains papers presented at the 4th *International Conference on Computer Science, Applied Mathematics and Applications* (ICCSAMA 2016) held on 2–3 May 2016 in Vienna, Austria. The conference is co-organized by International Institute for Applied Systems Analysis (IIASA), Austria, in cooperation with Department of Information Systems (Wrocław University of Technology, Poland), Laboratory of Theoretical and Applied Computer Science LITA (Lorraine University, France), Analysis, Design and Development of ICT systems Laboratory, Budapest University of Technology and Economics, Hungary and IEEE SMC Technical Committee on Computational Collective Intelligence.

The aim of ICCSAMA 2016 is to bring together leading academic scientists, researchers and scholars to discuss and share their newest results in the fields of computer science, applied mathematics and their applications. After the peer-review process, 20 papers by authors from Algeria, Austria, France, Germany, Greece, Georgia, Hungary, Italy, Malaysia, Spain, Turkey and Vietnam have been selected for including in these proceedings. The presentations of 19 have been partitioned into five sessions: *Advanced Optimization Methods and Their Applications, Models for ICT applications, Topics on discrete mathematics, Data Analytic Methods and Applications* and *Feature Extraction.*

The clear message of the proceedings is that the potentials of computational methods for knowledge engineering and optimization algorithms are to be exploited, and this is an opportunity and a challenge for researchers. It is observed that the ICCSAMA 2013, 2014 and 2015 clearly generated a significant amount of interaction between members of both computer science and applied mathematics communities. The intensive discussions have seeded future exciting development at the interface between computational methods, optimization and engineering.

The works included in these proceedings can be useful for researchers, Ph.D. and graduate students in optimization theory and knowledge engineering fields. It is the hope of the editors that readers can find many inspiring ideas and use them to their research. Many such challenges are suggested by particular approaches and models presented in the proceedings.

We would like to thank all authors, who contributed to the success of the conference and to this book. Special thanks go to the members of the Steering and Program Committees for their contributions to keeping the high quality of the selected papers. Cordial thanks are due to the Organizing Committee members for their efforts and the organizational work.

Finally, we cordially thank Springer for support and publishing this volume.

We hope that ICCSAMA 2016 significantly contributes to the fulfilment of the academic excellence and leads to greater success of ICCSAMA events in the future.

May 2016   Thanh Binh Nguyen
                                                                          Tien Van Do
                                                                       Hoai An Le Thi
                                                                   Ngoc Thanh Nguyen

ICCSAMA 2016 Organization

General Chair

Nguyen Thanh Binh, International Institute for Applied Systems Analysis (IIASA)
Nguyen Ngoc Thanh, Wrocław University of Technology, Poland

General Co-chairs

Le Thi Hoai An, Lorraine University, France
Tien Van Do, Budapest University of Technology and Economics, Hungary

Program Chairs

Pham Dinh Tao, INSA Rouen, France
Le Nguyen-Thinh, Humboldt-Universität zu Berlin, Germany

Doctoral Track Chair

Nguyen Anh Linh, Warsaw University, Poland

Organizing Committee

Marcos Carmen, International Institute for Applied Systems Analysis (IIASA)

Steering Committee

Nguyen Thanh Binh, International Institute for Applied Systems Analysis (IIASA), Austria (Co-chair)
Nguyen Ngoc Thanh, Wrocław University of Technology, Poland (Co-Chair)
Le Thi Hoai An, Lorraine University, France (Co-chair)
Tien Van Do, Budapest University of Technology and Economics, Hungary (Co-chair)
Pham Dinh Tao, INSA Rouen, France
Nguyen Hung Son, Warsaw University, Poland
Nguyen Anh Linh, Warsaw University, Poland
Tran Dinh Viet, Slovak Academy of Sciences, Slovakia

Program Committee

Attila Kiss, Eötvös Loránd University, Budapest
Bui Alain, Université de Versailles-St-Quentin-en-Yvelines, France
Bui Minh-Phong, Eötvös Loránd University, Budapest
Ha Quang Thuy, Vietnam National University, Vietnam
Le Nguyen-Thinh, Humboldt Universität zu Berlin, Germany
Le Thi Hoai An, Lorraine University, France
Ngo Van Sang, University of Rouen, France
Nguyen Anh Linh, Warsaw University, Poland
Nguyen Benjamin, University of Versailles Saint-Quentin-en-Yvelines, France
Nguyen Duc Cuong, International University VNU-HCM, Vietnam
Nguyen Hung Son, Warsaw University, Poland
Nguyen Ngoc Thanh, Wrocław University of Technology, Poland
Nguyen Thanh Binh, International Institute for Applied Systems Analysis (IIASA), Austria
Nguyen Viet Hung, Laboratory of Computer Sciences Paris 6, France
Pham Cong Duc, University of Pau and Pays de l'Adour, France
Pham Dinh Tao, INSA Rouen, France
Sztrik János, Debrecen University, Hungary
Tien Van Do, Budapest University of Technology and Economics, Hungary
Tran Dinh Viet, Slovak Academy of Sciences, Slovakia
Tran Quoc-Binh, Debrecen University, Hungary

Contents

Part I
Advanced Optimization Methods and Their Applications

Part I
Advanced Optimization Methods
and Their Application

A DC Programming Approach to the Continuous Equilibrium Network Design Problem

Thi Minh Tam Nguyen and Hoai An Le Thi

Abstract In this paper, we consider one of the most challenging problems in transportation, namely the Continuous Equilibrium Network Design Problem (CENDP). This problem is to determine capacity expansions of existing links in order to minimize the total travel cost plus the investment cost for link capacity improvements, when the link flows are constrained to be in equilibrium. We used the model of mathematical programming with complementarity constraints (MPCC) for the CENDP and recast it as a DC (Difference of Convex functions) program with DC constraints via the use of a penalty technique. A DCA (DC Algorithm) was developed to solve the resulting problem. Numerical results indicate the efficiency of our method vis-à-vis some existing algorithms.

Keywords CENDP · MPCC · Complementarity constraint · DC programming · DCA · DC constraints · Penalty technique

1 Introduction

The network design problem is one of the critical problems in transportation due to increasing demand for travel on roads. The purpose of this problem is to select location to build new links or to determine capacity improvements of existing links so as to optimize transportation network in some sense. A network design problem is said to be continuous if it deals with divisible capacity expansions (expressed by continuous variables). The CENDP consists of determining capacity enhancement of existing links to minimize the sum of the total travel cost and the expenditure for link capacity improvement, when the link flows are restricted to be in equilibrium.

T.M.T. Nguyen (✉) · H.A. Le Thi
Laboratory of Theoretical and Applied Computer Science (LITA),
UFR MIM University of Lorraine, Ile du Saulcy, 57045 Metz, France
e-mail: thi-minh-tam.nguyen@univ-lorraine.fr

H.A. Le Thi
e-mail: hoai-an.le-thi@univ-lorraine.fr

© Springer International Publishing Switzerland 2016
T.B. Nguyen et al. (eds.), *Advanced Computational Methods
for Knowledge Engineering*, Advances in Intelligent Systems
and Computing 453, DOI 10.1007/978-3-319-38884-7_1

3

The term "equilibrium" in this problem refers to the deterministic user equilibrium which is defined that for each origin-destination pair, at equilibrium, the travel costs on all utilized paths equal and do not exceed the travel cost on any unused path. The CENDP is generally formulated as a bi-level program or mathematical program with equilibrium constraints. In general, solving these problems is intractable because of the nonconvexity of both the objective function and feasible region.

Abdulaal and LeBlanc [1] are presumed to be pioneers studying the CENDP. They stated this problem as a bi-level program and transformed it into an unconstrained problem which was solved by the Hooke-Jeeves' algorithm. To date, a number of approaches have been proposed to address this CENDP, for example the equilibrium decomposed optimization heuristics [13], sensitivity analysis based heuristic methods [4], simulated annealing approach [3], augmented Lagrangian method [10], gradient-based approaches [2]. More recently, Wang et al. [14] suggested a method for finding the global solution to the linearized CENDP. These authors formulated the CENDP as an MPCC and converted complementarity constraints into the mixed-integer linear constraints. In addition, the travel cost functions were linearized by introducing binary variables. As a result, the MPCC became a mixed-integer linear program that was solved by using optimization software package CPLEX. However, their method for solving the MPCC produces many new variables including a considerable number of the binary variables and the mixed-integer linear program itself is a hard problem.

The aim of this paper is to give a new approach based on DC programming and DCA to solve the MPCC model for the CENDP. DC programming and DCA was introduced by Pham Dinh Tao in 1985 in their preliminary form and extensively developed by Le Thi Hoai An and Pham Dinh Tao since 1994 ([8, 11, 12] and the references therein). They are classic now and used by plenty of researchers in various fields ([5, 15, 16] and the list of references in [6]). Although DCA is a local optimization approach, it provides quite often a global solution and is proved be more robust and efficient than the standard methods. The success of DCA in many studies motivated us to investigate it for addressing the CENDP.

The rest of the paper is organized as follows. In Sect. 2, we describe the MPCC model for the CENDP. Section 3 presents a brief introduction of DC programming and DCA and the solution method for the CENDP. The numerical results are reported in Sect. 4 and some conclusions are given in Sect. 5.

2   Problem Formulation

The following notation is used throughout this paper.

A	the set of links in the network.
W	the set of origin-destination (O-D) pairs.
R_w	the set of paths connecting the O-D pair $w \in W$.
q_w	the fixed travel demand for O-D pair w.

δ_{ap}^w the indicator variables, $\delta_{ap}^w = 1$ if link a is on path p between O-D pair w, $\delta_{ap}^w = 0$ otherwise.

f_p^w the flow on path p connecting O-D pair w, $f = [f_p^w]$.

x_a the flow on link a, $x = [x_a]$,

$$x_a = \sum_{w \in W} \sum_{p \in R_w} \delta_{ap}^w f_p^w.$$

y_a the capacity of link a after expansion, $y = [y_a]$.

$\underline{y_a}$ the capacity of link a before expansion, $\underline{y} = [\underline{y_a}]$.

$\overline{y_a}$ the upper bound of y_a.

π_w the minimum travel cost between O-D pair w.

$g_a(y_a)$ the improvement cost for link a.

θ the relative weight of improvement costs and travel costs.

$t_a(x_a, y_a)$ the travel cost on link a,

$$t_a(x_a, y_a) = A_a + B_a \left(\frac{x_a}{y_a} \right)^4.$$

c_p^w the travel cost on path p between O-D pair w, $c = [c_p^w]$,

$$c_p^w = \sum_{a \in A} \delta_{ap}^w t_a(x_a, y_a).$$

In this paper, we assume that g_a is convex.

As mentioned in [14], the CENDP can be formulated as the following MPCC:

$$\min_{x,y,f,c,\pi} \sum_{a \in A} t_a(x_a, y_a) x_a + \theta \sum_{a \in A} g_a(y_a) \quad \text{(MPCC)}$$

subject to:

(i) Demand conservation and capacity expansion constraints:

$$\sum_{p \in R_w} f_p^w = q_w, \ w \in W, \tag{1}$$

$$\underline{y_a} \leq y_a \leq \overline{y_a}, \ a \in A. \tag{2}$$

(ii) Deterministic user equilibrium constraints:

$$f_p^w(c_p^w - \pi_w) = 0, \ p \in R_w, w \in W, \tag{3}$$

$$c_p^w - \pi_w \geq 0, \ p \in R_w, w \in W. \tag{4}$$

(iii) Definitional constraints:

$$x_a = \sum_{w \in W} \sum_{p \in R_w} \delta_{ap}^w f_p^w, \ x_a \geq 0, \ a \in A, \tag{5}$$

$$c_p^w = \sum_{a \in A} \delta_{ap}^w \left[A_a + B_a \left(\frac{x_a}{y_a} \right)^4 \right], \ p \in R_w, w \in W, \tag{6}$$

$$f_p^w \geq 0, \ p \in R_w, w \in W. \tag{7}$$

3   Solution Method by DC Programming and DCA

Before presenting the DCA for solving the problem (MPCC), we introduce briefly DC programming and DCA.

3.1   Introduction to DC Programming and DCA

DC Programming and DCA constitute the backbone of smooth/nonsmooth noncon-vex programming and global optimization. They address the problem of minimizing a function f which is a difference of convex functions on the whole space $\mathbb{R}^n$ or on a convex set $C \subset \mathbb{R}^n$. Generally speaking, a standard DC program takes the form

$$\alpha = \inf \{f(x) := g(x) - h(x) : \ x \in \mathbb{R}^n\} \tag{8}$$

where $g, h \in \Gamma_0(\mathbb{R}^n)$, the set of lower semi-continuous proper convex functions on $\mathbb{R}^n$. Such a function f is called a DC function, and $g - h$, a DC decomposition of f, while the convex functions g and h are DC components of f. The convex constraint $x \in C$ can be incorporated in the objective function of (8) by using the indicator function on C.

For a convex function ϕ, the subdifferential of ϕ at x_0, denoted as $\partial \phi(x_0)$, is defined by

$$\partial \phi(x_0) := \{y \in \mathbb{R}^n : \ \phi(x) \geq \phi(x_0) + \langle x - x_0, y \rangle, \forall x \in \mathbb{R}^n\}.$$

The idea of DCA for solving the problem (8) is quite simple: each iteration k of DCA approximates the concave part $-h$ by its affine majorization (that corresponds to taking $y^k \in \partial h(x^k)$) and minimizes the resulting convex function. For a complete study of standard DC programming and DCA, readers are referred to [8, 11] and the references therein.

Recently, Le Thi et al. [7, 12] investigated the extension of DC programming and DCA to solve DC programs with DC constraints which are of the form

$$
\begin{aligned}
\min \ &g(x) - h(x) \\
\text{s.t.} \ &x \in C, \\
&g_i(x) - h_i(x) \le 0, i = 1, \dots, m,
\end{aligned}
\tag{9}
$$

where $C \subseteq \mathbb{R}^n$ is a nonempty closed convex set; $g, h, g_i, h_i \in \Gamma_0(\mathbb{R}^n), i = 1, \dots, m$.

This class of nonconvex programs is the most general in DC programming and solving them is more difficult than solving standard DC programs because of the nonconvexity of the constraints. Two approaches for the problem (9) were proposed in [7]. The first one is based on penalty techniques in DC programming while the second one linearizes concave parts in DC constraints to build convex inner approximations of the feasible set. Since we use the second approach to solve the CENDP, we shortly present herein this approach for the problem (9).

Using the main idea of the DCA that linearizes the concave part of the DC structure, we can derive a sequential convex programming method based on solving the following convex subproblems

$$
\begin{aligned}
\min \ &g(x) - \langle y^k, x \rangle \\
\text{s.t.} \ &x \in C, \\
&g_i(x) - h_i(x^k) - \langle y_i^k, x - x^k \rangle \le 0, i = 1, \dots, m,
\end{aligned}
\tag{10}
$$

where $x^k \in \mathbb{R}^n$ is the current iterate, $y^k \in \partial h(x^k), y_i^k \in \partial h_i(x^k)$ for $i = 1, \dots, m$.

This linearization introduces an inner convex approximation of the feasible set of the problem (9). However, this convex approximation is quite often poor and can lead to infeasibility of convex subproblem (10). To deal with the feasibility of subproblems, a relaxation technique was proposed. Instead of (10), we consider the subproblem

$$
\begin{aligned}
\min \ &g(x) - \langle y^k, x \rangle + t_k s \\
\text{s.t.} \ &x \in C, \\
&g_i(x) - h_i(x^k) - \langle y_i^k, x - x^k \rangle \le s, i = 1, \dots, m, \\
&s \ge 0,
\end{aligned}
\tag{11}
$$

where $t_k > 0$ is a penalty parameter. Clearly, (11) is a convex problem that is always feasible. Moreover, the Slater constraint qualification is satisfied for the constraints of (11), thus the Karush-Kuhn-Tucker (KKT) optimality condition holds for some solution (x^{k+1}, s^{k+1}) of (11).

DCA scheme for the DC program with DC constraints (9)

- **Initialization**. Choose an initial point x^0; $\delta_1, \delta_2 > 0$, an initial penalty parameter $t_1 > 0$. Set $0 \longleftarrow k$.
- **Repeat**.
  Step 1. Compute $y^k \in \partial h(x^k)$, $y_i^k \in \partial h_i(x^k)$, $i = 1, \dots, m$,
  Step 2. Compute (x^{k+1}, s^{k+1}) as the solution of the convex problem (11), and the associated Lagrange multipliers $(\lambda^{k+1}, \mu^{k+1})$,
  Step 3. Penalty parameter update.
  compute $r_k = \min\{\|x^{k+1} - x^k\|^{-1}, \|\lambda^{k+1}\|_1 + \delta_1\}$
  and set $\beta_{k+1} = \begin{cases} \beta_k & \text{if } \beta_k \geq r_k, \\ \beta_k + \delta_2 & \text{if } \beta_k < r_k. \end{cases}$
  Step 4. $k \leftarrow k + 1$,
- **Until** $x^{k+1} = x^k$ and $s^{k+1} = 0$.

The global convergence of the above algorithm is completely proved in [7].

3.2   DCA for Solving the Problem (MPCC)

It is worth noting that

$$\sum_{a\in A} t_a(x_a, y_a)x_a = \sum_{w\in W} q_w \pi_w,$$

therefore the objective function of the problem (MPCC) is equal to

$$\sum_{w\in W} q_w \pi_w + \theta \sum_{a\in A} g_a(y_a)$$

which is convex. However, the MPCC is still a difficult problem due to the non-convexity of the feasible region which stems from complementarity constraints and the non-linear travel cost functions. To handle this problem, the complementarity constraints

$$\left\{ f_p^w(c_p^w - \pi_w) = 0, f_p^w \geq 0, c_p^w - \pi_w \geq 0 \right\}$$

are replaced by

$$\left\{ \min(f_p^w, c_p^w - \pi_w) \leq 0, f_p^w \geq 0, c_p^w - \pi_w \geq 0 \right\}.$$

Besides, for $a \in A$, the new variables $u_a = \frac{x_a^2}{y_a^2}$ are introduced to lessen the level of complexity of the constraints (6).

The (MPCC) can be rewritten as follows:

$$\min_{x,y,f,c,\pi,u} \sum_{w\in W} q_w \pi_w + \theta \sum_{a\in A} g_a(y_a)$$

s.t. (1), (2), (3), (4), (6),

$$\min(f_p^w, c_p^w - \pi_w) \leq 0, \ p \in R_w, w \in W,$$

$$c_p^w - \sum_{a\in A} \delta_{ap}^w (A_a + B_a u_a^2) = 0, \ p \in R_w, w \in W,$$

$$\frac{x_a^2}{y_a} - u_a y_a = 0, \ a \in A.$$

This problem is equivalent to the following problem:

$$\min_{x,y,f,c,\pi,u,v} \sum_{w\in W} q_w \pi_w + \theta \sum_{a\in A} g_a(y_a) \quad (P1)$$

s.t. (1), (2), (3), (4), (6),

$$\min(f_p^w, c_p^w - \pi_w) \leq 0, \ p \in R_w, w \in W, \tag{12}$$

$$c_p^w - \sum_{a\in A} \delta_{ap}^w (A_a + B_a u_a^2) \leq 0, \ p \in R_w, w \in W, \tag{13}$$

$$\sum_{a\in A} \delta_{ap}^w (A_a + B_a u_a^2) - c_p^w \leq 0, \ p \in R_w, w \in W, \tag{14}$$

$$\frac{x_a^2}{y_a} - v_a \leq 0, \ a \in A, \tag{15}$$

$$v_a - u_a y_a \leq 0, \ a \in A, \tag{16}$$

$$u_a y_a - \frac{x_a^2}{y_a} \leq 0, \ a \in A. \tag{17}$$

The problem (P1) can be solved by transforming the constraints (12), (13) and (15)–(17) into DC constraints and developing a DCA for the resulting problem. However, when the total number of paths is large this method produces many DC constraints. To diminish the number of these DC constraints, the constraints (12), (13) of the problem (P1) are penalized and we obtain the problem:

$$\min \sum_{w\in W} q_w \pi_w + \theta \sum_{a\in A} g_a(y_a) + t_1 F_1(X) + t_2 F_2(X) \quad (P2)$$

s.t. $X \in P,$

$$\sum_{a\in A} \delta_{ap}^w (A_a + B_a u_a^2) - c_p^w \leq 0, \ p \in R_w, w \in W,$$

$$x_a^2 - v_a y_a \leq 0, \ a \in A,$$

$$v_a - u_a y_a \leq 0, \ a \in A,$$

$$u_a y_a - \frac{x_a^2}{y_a} \leq 0, \ a \in A,$$

where $F_1(X) = \sum_{p,w} \min(f_p^w, c_p^w - \pi_w), F_2(X) = \sum_{p,w} \left(c_p^w - \sum_{a \in A} \delta_{ap}^w (A_a + B_a u_a^2) \right),$

$P = \{ X = (x, y, f, c, \pi, u, v) \mid X \text{ satisfies } (1), (2), (3), (4), (6) \text{ and } u_a \ge 0, \forall a \}.$

Obviously, P is a polyhedral convex set in $\mathbb{R}^n$ with $n = 4nL + 2nP + nOD$ (nL is the number of links, nP is the total number of paths, nOD is the number of O-D pairs).

It is easy to prove that if an optimal solution X^* to (P2) satisfies $F_1(X^*) = 0$ and $F_2(X^*) = 0$ then it is an optimal solution to the original problem. Furthermore, according to the general result of the penalty method (see [9], pp. 402–405), for large numbers t_1, t_2, the minimum point of problem (P2) will be in a region where F_1, F_2 are relatively small. For this reason, we consider the penalized problem (P2) with sufficiently large values of t_1, t_2.

Using the equalities

$$MN = \frac{1}{2}(M + N)^2 - \frac{1}{2}\left(M^2 + N^2\right) = \frac{1}{2}\left(M^2 + N^2\right) - \frac{1}{2}(M - N)^2,$$

the penalized problem (P2) can be reformulated as the DC program with DC constraints:

$$\min \ G(X) - H(X) \quad \text{(P3)}$$
$$\text{s.t. } X \in P,$$
$$\sum_{a \in A} \delta_{ap}^w (A_a + B_a u_a^2) - c_p^w \le 0, \ p \in R_w, w \in W,$$
$$G_{ia}(X) - H_{ia}(X) \le 0, \ i = 1, 2, 3; \ a \in A,$$

where $\quad G(X) = \sum_{w \in W} q_w \pi_w + \theta \sum_{a \in A} g_a(y_a) + t_2 \sum_{p,w} \left(c_p^w - \sum_{a \in A} \delta_{ap}^w A_a \right),$

$$H(X) = t_1 \sum_{p,w} \max(-f_p^w, -c_p^w + \pi_w) + t_2 \sum_{p,w} \left(\sum_{a \in A} \delta_{ap}^w B_a u_a^2 \right),$$

$$G_{1a}(X) = x_a^2 + \frac{1}{2}\left(v_a^2 + y_a^2\right), \ H_{1a}(X) = \frac{1}{2}(v_a + y_a)^2,$$

$$G_{2a}(X) = v_a + \frac{1}{2}\left(u_a^2 + y_a^2\right), \ H_{2a}(X) = \frac{1}{2}(u_a + y_a)^2,$$

$$G_{3a}(X) = \frac{1}{2}(u_a^2 + y_a^2), \ H_{3a}(X) = \frac{x_a^2}{y_a} + \frac{1}{2}(u_a - y_a)^2.$$

Adapting the generic DCA scheme for DC programs with DC constraints, we propose a DCA for solving the problem (P3). At each iteration k, we compute

$$Y^k \in \partial H(X^k), \ Y_{ia}^k \in \partial H_{ia}(X^k), \ i = 1, 2, 3; \ a = 1, \ldots, nL$$

and then solve the following convex problem:

$$\min \ F(X, s) = G(X) - \langle Y^k, X \rangle + ts$$

$$\text{s.t. } X \in P,$$

$$\sum_{a \in A} \delta_{ap}^w (A_a + B_a u_a^2) - c_p^w \le 0, \ p \in R_w, w \in W, \tag{18}$$

$$G_{ia}(X) - H_{ia}(X^k) - \langle Y_{ia}^k, X - X^k \rangle \le s, \ i = 1, 2, 3; \ a = 1, \dots, nL,$$

$$s \ge 0.$$

A subgradient $Y = (\widehat{x}, \widehat{y}, \widehat{f}, \widehat{c}, \widehat{\pi}, \widehat{u}, \widehat{v}) \in \partial H(X)$ can be selected as follows:

$$\widehat{x} = \widehat{y} = 0, \ \widehat{f}_p^w = \begin{cases} -t_1 & \text{if } f_p^w < c_p^w - \pi_w \\ 0 & \text{otherwise} \end{cases}, \ \widehat{c}_p^w = \begin{cases} 0 & \text{if } f_p^w < c_p^w - \pi_w \\ -t_1 & \text{otherwise} \end{cases} \tag{19}$$

$$\widehat{\pi}_w = \begin{cases} 0 & \text{if } f_p^w < c_p^w - \pi_w \\ t_1 & \text{otherwise} \end{cases}, \ \widehat{u}_a = 2t_2 \sum_{p,w} \delta_{ap}^w B_a u_a, \ \widehat{v} = 0. \tag{20}$$

Since H_{ia} is differentiable, $\partial H_{ia}(X) = \{\nabla H_{ia}(X)\}$, $i = 1, 2, 3$; $a = 1, \dots, nL$. Consider $Y = (\widehat{x}, \widehat{y}, \widehat{f}, \widehat{c}, \widehat{\pi}, \widehat{u}, \widehat{v})$, we have

$$Y = \nabla H_{1a}(X) \Leftrightarrow \widehat{x} = \widehat{f} = \widehat{c} = \widehat{\pi} = \widehat{u} = 0, \ \widehat{y}_a = \widehat{v}_a = v_a + y_a, \tag{21}$$

$$Y = \nabla H_{2a}(X) \Leftrightarrow \widehat{x} = \widehat{f} = \widehat{c} = \widehat{\pi} = \widehat{v} = 0, \ \widehat{y}_a = \widehat{u}_a = u_a + y_a, \tag{22}$$

$$Y = \nabla H_{3a}(X)$$

$$\Leftrightarrow \widehat{x}_a = \frac{2x_a}{y_a}, \widehat{y}_a = -\frac{x_a^2}{y_a^2} + y_a - u_a, \widehat{f} = \widehat{c} = \widehat{\pi} = \widehat{v} = 0, \widehat{u}_a = u_a - y_a. \tag{23}$$

Algorithm (DCA for the problem (P3))

- **Initialization.** Choose an initial point $X^0 \in \mathbb{R}^n$ and a penalty parameter $t = t_0$. Set $k = 0$ and let $\varepsilon_1, \varepsilon_2$ be sufficiently small positive numbers.
- **Repeat.**
  Step 1. Compute $Y^k \in \partial H(X^k)$, $Y_{ia}^k \in \partial H_{ia}(X^k)$, $i = 1, 2, 3$; $a = 1, \dots, nL$ via (19)–(23).
  Step 2. Compute (X^{k+1}, s^{k+1}) as the solution of (18).
  Step 3. $k \leftarrow k + 1$
- **Until** $s \le \varepsilon_2$ and $(\|X^{k+1} - X^k\| \le \varepsilon_1(\|X^k\| + 1))$ or $|F(X^{k+1}, s^{k+1}) - F(X^k, s^k)| \le \varepsilon_1(|F(X^k, s^k)| + 1))$.

4   Numerical Results

In order to illustrate the efficiency of the proposed algorithm, numerical experiments
were performed on a network consisting of 6 nodes and 16 links (Fig. 1). This network has been used to test the different algorithms that deal with the problem CENDP
[2, 3, 10, 13, 14]. The travel demand for this network was considered in two cases
(given in Table 1). The detailed data including the parameters of travel cost functions and enhancement cost functions, the capacity of links before expansion, the
upper bound of the capacity improvements can be found in [3]. Our algorithm was
compared with eight other methods. The abbreviations for these methods and their
sources are listed in Table 2. We also made a comparison with a good lower bound
of the CENDP (mentioned in [2] and labeled as "SO").

Fig. 1   16-link network

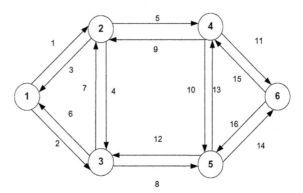

Table 1   Level of travel demand

	Case I	Case II
Demand from node 1 to node 6	5	10
Demand from node 6 to node 1	10	20
Total travel demand	15	30

Table 2   Abbreviation of method names

Abbreviation	Name	Source
EDO	Equilibrium decomposed optimization	[13]
SA	Simulated annealing algorithm	[3]
AL	Augmented lagrangian algorithm	[10]
GP	Gradient projection method	[2]
CG	Conjugate gradient projection method	[2]
QNEW	Quasi-NEWton projection method	[2]
PT	PARATAN version of gradient projection method	[2]
MILP	Mixed-Integer linear program transformation	[14]
DCA	DC algorithm	This paper

Table 3 The numerical results of DCA and the existing algorithms in case I

	EDO	SA	AL	GP	CQ	QNEW	PT	MILP	DCA	SO
$\triangle y_1$	0	0	0	0	0	0	0	0	0	0
$\triangle y_2$	0	0	0	0	0	0	0	0	0	0
$\triangle y_3$	0.13	0	0.0062	0	0	0	0	0	0	0
$\triangle y_4$	0	0	0	0	0	0	0	0	0	0
$\triangle y_5$	0	0	0	0	0	0	0	0	0	0
$\triangle y_6$	6.26	3.1639	5.2631	5.8302	6.1989	6.0021	5.9502	4.41	4.8487	5.9979
$\triangle y_7$	0	0	0.0032	0	0	0	0	0	0	0
$\triangle y_8$	0	0	0	0	0	0	0	0	0	0
$\triangle y_9$	0	0	0	0	0	0	0	0	0	0
$\triangle y_{10}$	0	0	0	0	0	0	0	0	0	0
$\triangle y_{11}$	0	0	0.0064	0	0	0	0	0	0	0
$\triangle y_{12}$	0	0	0	0	0	0	0	0	0	0
$\triangle y_{13}$	0	0	0	0	0	0	0	0	0	0
$\triangle y_{14}$	0	0	0	0	0	0	0	0	0	0
$\triangle y_{15}$	0.13	0	0.7171	0.87	0.0849	0.1846	0.5798	0	0	0.1449
$\triangle y_{16}$	6.26	6.7240	6.7561	6.1090	7.5888	7.5438	7.1064	7.7	7.5621	7.5443
Obj	201.84	**198.104**	202.991	202.24	199.27	198.68	200.60	199.781	**199.656**	193.39
Gap	4.37	2.44	4.96	4.58	3.04	2.74	3.73	3.30	3.24	0

Table 4 The numerical results of DCA and the existing algorithms in case II

	EDO	SA	AL	GP	CQ	QNEW	PT	MILP	DCA	SO
$\triangle y_1$	0	0	0	0.1013	0.1022	0.0916	0.101	0	0	0.1165
$\triangle y_2$	4.88	0	4.6153	2.1818	2.1796	2.1521	2.1801	4.41	4.6144	2.1467
$\triangle y_3$	8.59	10.1740	9.8804	9.3423	9.3425	9.1408	9.3339	10.00	9.9166	9.3447
$\triangle y_4$	0	0	0	0	0	0	0	0	0	0
$\triangle y_5$	0	0	0	0	0	0	0	0	0	0
$\triangle y_6$	7.48	5.7769	7.5995	9.0443	9.0441	8.8503	9.0361	7.42	7.3442	9.0424
$\triangle y_7$	0.26	0	0.0016	0	0	0	0	0	0	0
$\triangle y_8$	0.85	0	0.6001	0.008	0.0074	0.0114	0.0079	0.54	0.5922	0
$\triangle y_9$	0	0	0.001	0	0	0	0	0	0	0
$\triangle y_{10}$	0	0	0	0	0	0	0	0	0	0
$\triangle y_{11}$	0	0	0	0	0	0	0	0	0	0
$\triangle y_{12}$	0	0	0.1130	0.0375	0.0358	0.0377	0	0	0	0
$\triangle y_{13}$	0	0	0	0	0	0	0	0	0	0
$\triangle y_{14}$	1.54	0	1.3184	0.0089	0.0083	0.0129	0.0089	1.18	1.3153	0
$\triangle y_{15}$	0.26	0	2.7265	1.9433	1.9483	1.9706	1.9429	0	0	1.7995
$\triangle y_{16}$	12.52	17.2786	17.5774	18.9859	18.986	18.575	18.9687	19.50	20	18.9875
Obj	540.74	528.497	532.71	534.017	534.109	534.08	534.02	523.627	**522.644**	512.013
Gap	5.61	3.22	4.04	4.30	4.32	4.31	4.30	2.27	2.08	0

In the DCA for the CENDP, we took $\varepsilon_1 = \varepsilon_2 = 10^{-6}$, the initial point was selected as follows:

$$X^0 = (x^0, y^0, f^0, c^0, \pi^0, u^0, v^0)$$

with

$$x^0 = y^0 = f^0 = u^0 = v^0 = 0, \quad c_p^{0w} = \sum_{a \in A} \delta_{ap}^w A_a, \quad \pi_w^0 = \min \left\{ c_p^{0w} : p \in R_w \right\}$$

The penalty parameters were chosen to be $t_1 = 350, t_2 = 100, t = 500$ and $t_1 = 400, t_2 = 100, t = 450$ for the cases I and II respectively.

For the MILP approach, the feasible region of link flow (resp. the feasible domain of the road capacity expansion) was divided into 10 equal segments. According to the authors, this partition is quite fine.

The computational results for the cases I and II are summarized in Tables 3 and 4 respectively. These results are taken from the previous work [2, 14] except for the results of DCA. In Tables 3 and 4, we use the following notations:

- $\triangle y_a$: the capacity enhancement on link a.
- Obj: the value of objective function.
- Gap: the relative difference with respect to SO defined by

$$\text{Gap}(\%) = \frac{100(\text{Obj} - \text{LB})}{\text{LB}}$$

where LB is the lower bound of the CENDP.

From Tables 3 and 4, we observe that the optimal value provided by DCA is quite close to the lower bound of the CENDP (the relative difference is 3.24 and 2.08 % in the cases I and II respectively). In the case I with the lower travel demand, DCA outperforms 5 out of 8 algorithms. Although the optimal value given by DCA is not better than those given by SA, CQ and QNEW, the relative difference between this optimal value and the best one (found by SA, a global optimization technique) is fairly small (0.78 %). In the case II, when the travel demand increases, DCA yields the optimal value that is superior to those computed by all the other methods.

5 Conclusions

We proposed an algorithm based on DC programming and DCA for handling the CENDP. We considered the MPCC model for this problem and employed a penaly technique to reformulate it as a DC program with DC constraints. Numerical experiments on a small network indicate that DCA solves the CENDP more effectively than the existing methods. Besides, the optimal value provided by DCA is quite close to the lower bound of the CENDP. This shows that the solution found by DCA may be global, even though DCA is a local optimization approach.

Acknowledgments This research is funded by Foundation for Science and Technology Development of Ton Duc Thang University (FOSTECT), website: http://fostect.tdt.edu.vn, under Grant FOSTECT.2015.BR.15.

References

1. Abdulaal, M., LeBlanc, L.J.: Continuous equilibrium network design models. Transp. Res. Part B **13**, 19–32 (1979)
2. Chiou, S.W.: Bilevel programming for the continuous transport network design problem. Transp. Res. Part B **39**, 361–383 (2005)
3. Friesz, T.L., Cho, H.J., Mehta, N.J., Tobin, R.L., Anandalingam, G.: A simulated annealing approach to the network design problem with variational inequality constraints. Transp. Sci. **26**(1), 18–26 (1992)
4. Friesz, T.L., Tobin, R.L., Cho, H.J., Mehta, N.J.: Sensitivity analysis based heuristic algorithms for mathematical programs with variational inequality constraints. Math. Program. **48**, 265–284 (1990)
5. Hoang, N.T.: Linear convergence of a type of iterative sequences in nonconvex quadratic programming. J. Math. Anal. Appl. **423**(2), 1311–1319 (2015)
6. Le Thi, H.A.: DC programming and DCA. http://www.lita.univ-lorraine.fr/~lethi
7. Le Thi, H.A., Huynh, V.N., Pham Dinh, T.: DC programming and DCA for general DC programs. In: Advances in Intelligent Systems and Computing, pp. 15–35. Springer (2014)
8. Le Thi, H.A., Pham Dinh, T.: The DC (difference of convex functions) programming and DCA revisited with DC models of real world nonconvex optimization problems. Ann. Oper. Res. **133**, 23–46 (2005)
9. Luenberger, D.G., Ye, Y.: Linear and Nonlinear Programming. Springer (2008)
10. Meng, Q., Yang, H., Bell, M.G.H.: An equivalent continuously differentiable model and a locally convergent algorithm for the continuous network design problem. Transp. Res. Part B **35**, 83–105 (2001)
11. Pham Dinh, T., Le Thi, H.A.: Convex analysis approach to D.C. programming: theory, algorithms and applications. Acta Math. Vietnam **22**(1), 289–355 (1997)
12. Pham Dinh, T., Le Thi, H.A.: Recent advances in DC programming and DCA. Trans. Comput. Collect. Intell. **8342**, 1–37 (2014)
13. Suwansirikul, C., Friesz, T.L., Tobin, R.L.: Equilibrium decomposed optimization: a heuristic for continuous equilibrium network design problem. Transp. Sci. **21**(4), 254–263 (1987)
14. Wang, D., Lo, H.K.: Global optimum of the linearized network design problem with equilibrium flows. Transp. Res. Part B **44**(4), 482–492 (2010)
15. Wu, C., Li, C., Long, Q.: A DC programming approach for sensor network localization with uncertainties in anchor positions. J. Ind. Manag. Optim. **10**(3), 817–826 (2014)
16. Yin, P., Lou, Y., He, Q., Xin, J.: Minimization of l_{1-2} for compressed sensing. SIAM J. Sci. Comput. **37**(1), A536–A563 (2015)

A Method for Reducing the Number of Support Vectors in Fuzzy Support Vector Machine

Nguyen Manh Cuong and Nguyen Van Thien

Abstract We offer an efficient method to reduce the number of support vectors for Fuzzy Support Vector Machine. Firstly, we consider the Fuzzy Support Vector Machine model which was proposed by Lin and Wang. For the reducing the number of support vectors, we apply the l_0 regularization term to the dual form of this model. The resulting optimization problem is non-smooth and non-convex. The l_0 is then replaced by an approximation function. An algorithm which is based on DC programming and DCA is then investigated to solve this problem. Numerical results on real-world datasets show the efficiency and the superiority of our method versus the standard algorithm on both support vector reduction and classification.

Keywords DC Programming and DCA · Fuzzy Support Vector Machine · Support vector reduction

1 Introduction

Support Vector Machine (SVM) is one of the most popular techniques for pattern classification and regression estimation problem. In recent years, this topic has attracted a great deal of interest in many disciplines. In the training phase, all data in the same class of training set are treated with the assumption that they have the same role. However, noise data or outliers are inevitable in the applications [24]. In some cases, each input point may not be fully assigned to one of the classes. In 2002, Lin and Wang [27] proposed Fuzzy Support Vector Machine (FSVM) in which a fuzzy membership is applied to each input point. This helps to reduce the influence of non-support vectors, noise and outliers on the optimal hyperplane.

N. Manh Cuong (✉) · N. Van Thien
Faculty of Information Technology, Hanoi University of Industry - Vietnam,
Hanoi, Vietnam
e-mail: manhcuong.haui@yahoo.com

N. Van Thien
e-mail: nvthien1970@gmail.com

© Springer International Publishing Switzerland 2016   17
T.B. Nguyen et al. (eds.), *Advanced Computational Methods
for Knowledge Engineering*, Advances in Intelligent Systems
and Computing 453, DOI 10.1007/978-3-319-38884-7_2

A challenge of Machine Learning is the handling of the input dataset with very large dimension [20]. In SVM and FSVM, the number of support vector which obtains from the training phase is usually large. This causes the slowness of classification phase. Many methods are proposed to treat this issue (see [11] and reference therein). Their target is how to reduce the number of support vectors but without loss of generalization performance [11].

One of the first research for support vector reduction in SVM was proposed by Burges in 1996 [3] which computes an estimation of discriminator function in terms of a reduced set of vectors. This reduced set method is believed to be computationally expensive although its improvements have been mentioned in a number of later researches (see [29] for more details). In 2002, Downs et al. [4] proposed a method in which the unnecessary support vectors are detected based on its linear dependence on the other support vectors in the feature space. However, the reduction is not obvious in order to avoid any possible decline in accuracy [11]. Later, an iterative approach was proposed in which the training phase is repeated on the nested subsets of training set until the accuracy of classifier is acceptable or the number of support vectors stops decreasing. Belong to this approach, it is possible to mention here the method of Li and Zang [26], Lee and Mangasarian [22] and the deeply researches of Huang and Lee [23] and Keerthi et al. [36]. In [22], the l_1 regularization term $\| \alpha \|_1$ is used as one of the factors for the support vector reduction purpose.

On the theoretical point of view, using l_0 regularization $\| \alpha \|_0$ is a natural way to reduce the number of support vectors. Unfortunately, the minimization of l_0 term is a non-smooth and non-convex optimization problem [18, 20]. In 2010, Huang et al. proposed Sparse Support Vector Classification which using the l_0 regularization term in the primal form of the Support Vector Classification problem. The Expectation Maximization (EM) algorithm is then used to find the optimal solution. Recently, some efficient algorithms based on DC Programming and DCA are presented in [15, 16, 18–20, 30]. But they are applied for the SVM, Multi-classes SVM and/or S3VM models. However, these algorithms are intensely suitable for the linear separable data.

In this paper, we consider the Fuzzy Support Vector Machine model which was proposed by Lin and Wang [27]. For the purpose of support vector reduction, the l_0 regularization term of α is added to the dual form of this model with a turning parameter. The resulting problem is a non-smooth, non-convex optimization problem. The l_0 is then approximated by the Concave approximation function [1] that results to a non-convex optimization problem. Our solution method is based on Difference of Convex functions (DC) programming and DC Algorithms (DCA) that were introduced by Pham Dinh Tao in their preliminary form in 1985 and have been extensively developed since 1994 by Le Thi Hoai An and Pham Dinh Tao and become now classic and more and more popular (see, e.g. [12–21, 30–33] and references therein).

Our works are motivated by many folds: firstly, whereas several support vector reduction methods for SVM have been proposed in the literature (see e.g. [1–11, 22, 23, 25, 26, 29, 35, 37]), there exist a few works on this issue for FSVM. Secondly, finding an efficient algorithm for solving the non-convex optimization problem remains an important topic in recent years. Finally, DCA has been successfully applied to many (smooth or non-smooth) large-scale non-convex programs in various domains of applied sciences, in particular in Machine Learning (see e.g. [12–21, 30–33] and references therein) for which they provided quite often a global solution and proved to be more robust and efficient than standard methods. Note that with suitable DC decompositions, DCA generates most standard algorithms in convex and non-convex optimization and the *EM* method in [1] can be considered as a particular case of DCA.

The rest of the paper is organized as follows. In Sect. 2, we present a brief overview of FSVM model. The problem of the support vector reduction for FSVM and its approximation problem are described in Sect. 3. Section 4 is devoted to give a brief presentation of DC programming and DCA. A DC formulation and a DC algorithm for the considered problem are presented in the Sect. 5. Computational experiments are reported in Sect. 6 and finally, Sect. 7 concludes the paper.

2 Two Classes Fuzzy Support Vector Machine

First, let us present a brief overview on two classes SVM model. Let $\mathcal{X} = \{x_1, x_2, ..., x_n\}$ be a set of n vectors in $\mathbb{R}^d$ and $\mathcal{Y} = \{-1, 1\}$ be a set of class labels. Given a training set $X = \{(x_1, y_1), (x_2, y_2), ..., (x_n, y_n)\} \in \mathbb{R}^{n*(d+1)}$ where $x_i \in \mathcal{X}, y_i \in \mathcal{Y}, i = \{1, ..., n\}$.

Let f be a hyperplane to discriminate between data points with negative and positive labels. So, f is defined as follows:

$$f = \langle w^T, x \rangle + b \tag{1}$$

where $w \in \mathbb{R}^d$ and $b \in \mathbb{R}$. The goal is to determine the most appropriate hyperplane that separates the training set in the best way. So, the SVM model is defined as follows:

$$\min_{w,b,\xi} \left\{ C \sum_{i=1}^{n} \xi_i + \|w\|_2^2 \right\}, \tag{2}$$

subject to:

$$\Omega : \begin{cases} (w, b, \xi) \in \mathbb{R}^d \times \mathbb{R} \times \mathbb{R}_+^n : \\ y_i(\langle w, \phi(x_i) \rangle + b) \geq 1 - \xi_i, \forall 1 \leq i \leq n \end{cases}$$

where $\phi(.)$ is a non-linear function which transforms the input space to a higher dimensional space to provide a better class separation and $\langle ., . \rangle$ is the scalar product in $\mathbb{R}^d$ space. In the objective function, $\xi_i \in \mathbb{R}$ are slack variables and $C \sum_{i=1}^{n} \xi_i$ is the hinge loss term which presents the training classification errors. The remaining term is known as a regularization and C is a parameter that presents the trade-off between the hinge loss and the regularization term. The dual form of (2) is presented as follows:

$$\min_{\alpha} \frac{1}{2}\alpha^T \mathbb{H}\alpha - e^T\alpha \qquad (3)$$

subject to:

$$\Delta : \begin{cases} y^T\alpha = 0 \\ 0 \le \alpha_i \le C, i = 1, ..., n \end{cases}$$

where $\mathbb{H}$ is a symmetric matrix $\in \mathbb{R}^{n \times n}$ defined as follows:

$$\mathbb{H}_{i,j} = y_i y_j \langle \phi(x_i), \phi(x_j) \rangle = y_i y_j K(x_i, x_j). \qquad (4)$$

Here, K(.) is a kernel function which is defined as $K(x, y) = \langle \phi(x), \phi(y) \rangle$. Let α^*, b^* be an optimal solution of (3), the classifier now takes the form:

$$f(x) = \sum_{i=1}^{n} y_i \alpha_i^* K(x_i, x) + b^*. \qquad (5)$$

In 2001, Lin and Wang [11] proposed the Fuzzy Support Vector Machine model (FSVM). Let $X = \{(x_i, y_i, m_i), i = 1, ..., n\}$ be a training dataset. Here, $m_i \in \mathbb{R}$ be a fuzzy membership which is associated with training point (x_i, y_i). The model of fuzzy support vector machine is described as follows:

$$\min_{w,b,\xi} \left\{ C \sum_{i=1}^{n} m_i \xi_i + \|w\|_2^2 : (w, b, \xi) \in \Omega \right\}. \qquad (6)$$

In this model, each slack variable $\xi_i, i = 1, ..., n$ is associated with a corresponding fuzzy membership value m_i. The dual form of this problem is presented as follows:

$$\min_{\alpha} \frac{1}{2}\alpha^T \mathbb{H}\alpha - e^T\alpha \qquad (7)$$

subject to:

$$\Gamma : \begin{cases} y^T\alpha = 0 \\ 0 \le \alpha_i \le m_i C, i = 1, ..., n \end{cases}$$

There are several ways to generate the fuzzy membership values (see e.g. [34, 38]). Normally, these values are determined depending on the rough sets theory or the existence of outliers in data [34]. In 2012, Ling Jian and Zunquan Xia [28] proposed a simple way to define the fuzzy membership for each training point. Let x^+ and x^- be the positive and negative clusters. Let l^+ and l^- be respectively the number of points in each cluster and Φ be a mapping from the data space into the feature space. Let d_i be the distance between x_i^+ to the center of negative cluster:

$$d_i = \| \Phi(x_i^+) - \frac{1}{l^-} \sum_{j=1}^{l^-} \Phi(x_j^-) \|$$
$$= K(x_i^+, x_i^+) - \frac{2}{l^-} \sum_{j=1}^{l^-} K(x_i^+, x_j^-) + \frac{1}{(l^-)^2} \sum_{j=1}^{l^-} \sum_{k=1}^{l^-} K(x_j^-, x_k^-).$$

So, we can set the fuzzy membership of $x_j^-, j = 1, ..., l^-$ equal to 1 and the fuzzy membership of $x_i^+, i = 1, ..., l^+$ is calculated by:

$$m_i = \frac{d_i}{\max(d_i)}. \tag{8}$$

3   FSVM with l_0 Regularization

A natural way to reduce the number of support vectors is using the zero-norm $\|\alpha\|_0$. A minimization of this term provides a sparse solution of α and it is straightforward that l_0 norm should lead to a highly sparse model [10]. Therefore, we investigated a sparse FSVM model in which the l_0 regularization is added in the objective function of (7) with a turning parameter λ as follows:

$$\min_{\alpha \in \Gamma} \frac{1}{2} \alpha^T \mathbb{H}\alpha - e^T \alpha + \lambda \| \alpha \|_0. \tag{9}$$

Due to the l_0 term, the resulting problem is a non-smooth, non-convex optimization problem. Let $s : \mathbb{R} \to \mathbb{R}$ be a step function which defined as: $s(x) = 1$ for $x \neq 0$ and $s(x) = 0$ for $x = 0$. Then for $X \in \mathbb{R}^n$ we have $\|X\|_0 = \sum_{i=1}^{n} s(X_i)$. It is great difficult to optimize a noncontinuous problem [10]. So, the $s(x)$ then can be approximated by a continuous function φ. We first consider the piecewise exponential function which was presented in the study of Bradley and Mangasarian [1] by:

$$\varphi(x) = 1 - e^{-a|x|}, \tag{10}$$

where a is a positive parameter. So, we obtain the equivalent problem of (9) as follows:

$$\min_{\alpha \in \Gamma} \frac{1}{2} \alpha^T \mathbb{H}\alpha - e^T \alpha + \lambda \varphi(\alpha) \tag{11}$$

The resulting optimization is still non-convex. To solve this problem, our method is based on DC Programming and DCA which is presented in short in the next section.

4   An Overview of DC Programming and DCA

DC programming and DCA constitute the backbone of smooth/non-smooth non-convex programming and global optimization [31]. A general DC program takes the form:

$$\inf\{f(x) := G(x) - H(x) : x \in \mathbb{R}^p\}, \quad (P_{dc})$$

where G and H are lower semi-continuous proper convex functions on $\mathbb{R}^p$. Such a function f is called DC function, and $G - H$, DC decomposition of f while G and H are DC components of f. The convex constraint $x \in C$ can be incorporated in the objective function of (P_{dc}) by using the indicator function on C denoted χ_C which is defined by $\chi_C(x) = 0$ if $x \in C$; $+\infty$ otherwise.

The idea of DCA is simple: each iteration of DCA approximates the concave part $-H$ by its affine majorization (that corresponds to taking $y^k \in \partial H(x^k)$) and minimizes the resulting convex function. The generic DCA scheme can be described as follows:

Initialization: Let $x^0 \in \mathbb{R}^p$ be a best guess, $l \leftarrow 0$.
Repeat
- Calculate $y^l \in \partial H(x^l)$.
- Calculate $x^{l+1} \in \arg\min\{G(x) - H(x^l) - \langle x - x^l, y^l \rangle\} : x \in \mathbb{R}^p$.
- $l \leftarrow l + 1$.
Until convergence of $\{x^l\}$.

Convergence properties of DCA and its theoretical basis can be found in [31, 32], for instant it is important to mention that: DCA is a descent method (the sequences $\{G(x^k) - H(x^k)\}$ is decreasing) without linesearch. It has a linear convergence for general DC programs and a finite convergence for polyhedral DC programs (a DC program is called polyhedral DC program if either G or H is a polyhedral convex function [31]).

5   DC Formulation and DCA for Problem (11)

Firstly, the approximation function φ is expressed as a DC function as follows:

$$\varphi(x) = g(x) - h(x), \ x \in \mathbb{R} \tag{12}$$

where $g(x) = a|x|$ and $h(x) = a|x| - 1 + e^{-a|x|}$ are convex functions [17, 20]. Using this approximation, (11) can be expressed as follows:

$$\min_{\alpha \in \Gamma} \frac{1}{2}\alpha^T \mathbb{H}\alpha - e^T\alpha + \lambda g(\alpha) - \lambda h(\alpha). \tag{13}$$

Let $G(\alpha)$ and $H(\alpha)$ be the functions which are defined by:

$$G(\alpha) = \frac{1}{2}\alpha^T \mathbb{H}\alpha - e^T\alpha + \lambda g(\alpha) \text{ and } H(\alpha) = \lambda h(\alpha) \tag{14}$$

So, a DC formulation of (11) takes the form:

$$\min_{\alpha \in \Gamma} G(\alpha) - H(\alpha) \tag{15}$$

Since g and h are convex functions, so are G and H. Furthermore, since $g(x) = max(ax, -ax)$ is polyhedral convex, the function G is polyhedral convex and therefore, (15) is a polyhedral DC program [18, 20].

We now present the DC algorithm for (15). According to the DCA general scheme, at each iteration k, the DCA for (15) amounts to computing a sub-gradient $Y^k = (\overline{\alpha}^k)$ of H at $X^k = (\alpha^k)$ and then, solves the convex program:

$$\min G(\alpha) - \langle Y^k, X \rangle : X = (\alpha) \in \Gamma \tag{16}$$

to obtain X^{k+1}. The DCA for (15) is presented in details as follows:

FDPIE

Initialization Let ε be a tolerance sufficiently small, set $k = 0$.
Choose $X^0 = (\alpha^0)$ be a guess. Set a positive value for the parameter a.
Repeat
1. Compute $Y^k = (\overline{\alpha}^k) = \partial H(X^k) = \text{sgn}(\alpha^k)\lambda(a - ae^{-a|\alpha^k|})$.
2. Compute $X^{k+1} = (\alpha^{k+1})$ by solving the convex quadratic optimization problem
$$\min_{\alpha \in \Gamma} \frac{1}{2}\alpha^T \mathbb{H}\alpha - e^T\alpha + \lambda a|\alpha| - < \overline{\alpha}^k, \alpha > \tag{17}$$

3. $k \leftarrow k + 1$.
Until $\|X^{k+1} - X^k\| \le \varepsilon(1 + \|X^k\|)$.

Since (15) is a polyhedral DC program, FDPIE algorithm has a finite convergence [17]. Moreover, the solution furnished by FDPIE satisfies the necessary local optimality condition thanks to the differentiability of H [15].

6 Numerical Experiment

Our algorithm is compared with the standard algorithm for the support vector reduction in FSVM which applied the l_1 norm (FSVM-L1). In this situation, the convex optimization problem takes the form:

$$\min_{\alpha \in \Gamma} \frac{1}{2}\alpha^T \mathbb{H}\alpha - e^T\alpha + \lambda \| \alpha \|_1 \tag{18}$$

For solving this problem and the convex problems in FDPIE, we used the CPLEX solver version 11.2 and tested on the computer $Intel^R$ core I5 2×2.6 GHz, 4GB RAM. The algorithms are tested on the real-word datasets which obtained from

Table 1 The description of testing datasets

Dataset	#point	#attribute	#class
Thyroid	251	5	2
Breast cancer	263	9	2
Ionosphere	351	2	2
W60	569	30	2
Diabetis	768	8	2
German	1000	20	2
Splice	2991	60	2
Ringnorm	7400	20	2

IDA Benchmark Repository (http://mldata.org/repository/tags/data/IDA_Bench mark_Repository/) and UCI Machine Learning Repository (Breast Cancer, Ionosphere and W60). The datasets are presented in details in Table 1.

The parameters C and the scale σ of Gausian kernel are chosen by a five-folds cross-validation. For the approximation function φ, the parameter a is set to 1.5. The membership values m_i which are associated with training points x_i are calculated by (8). For each dataset, we applied the Boot Trap procedure to exact randomly 70 % of elements for training set and the remaining data for test set. This procedure is repeated 10 times and the average of the number of support vectors (res. the accuracy of classifiers) as well as its standard deviations are reported.

The turning parameter λ conducts the reducing of number of support vectors. Obviously, the larger value of λ is, the smaller number of support vectors would be. For evaluating the influent of λ to the sparsity of the solutions, in the experiment 1, the parameter λ is varied in $\{0.1, 0.2, ..., 0.9\}$ whereas all others parameters are fixed. The results on the Thyriod are reported in Fig. 1. In this figure, the percentage of the selected support vectors is calculated by dividing the number of selected support vectors for n which is considered as the maximum number of support vectors.

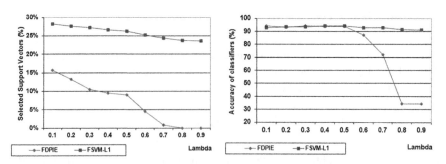

Fig. 1 Selected support vectors (*left*) and corresponding accuracy of classifiers (*right*) obtained from FDPIE and FSVM-L1 on Thyroid dataset

Table 2 Selected support vectors and accuracy of classifiers

Dataset	FDPIE		FSVM-L1		FSVM	
	#SV (%)	Acc (%)	#SV (%)	Acc (%)	#SV (%)	Acc (%)
Thyroid	9.03 ± 0.6	93.75 ± 1.97	26.22 ± 1.0	93.75 ± 1.51	38.73 ± 0.5	93.75 ± 2.96
Breast cancer	35.13 ± 0.3	72.31 ± 2.49	61.60 ± 6.8	72.30 ± 0.70	70.34 ± 0.0	68.46 ± 3.69
Ionosphere	12.25 ± 0.7	81.88 ± 1.65	30.48 ± 2.0	84.45 ± 0.72	39.72 ± 0.9	77.83 ± 0.71
W60	16.52 ± 0.0	91.01 ± 1.14	29.35 ± 0.4	94.71 ± 0.69	30.58 ± 0.8	88.88 ± 0.52
Diabetis	31.25 ± 1.6	72.61 ± 1.34	69.97 ± 0.1	72.61 ± 2.12	70.05 ± 0.0	72.09 ± 2.57
German	28.60 ± 0.3	71.53 ± 1.43	49.08 ± 1.9	69.20 ± 2.68	70.00 ± 0.0	69.13 ± 1.07
Splice	28.94 ± 0.5	73.85 ± 2.44	69.98 ± 0.0	73.54 ± 1.29	70.01 ± 0.0	73.33 ± 1.90
Ringnorm	11.18 ± 0.0	68.47 ± 0.65	64.49 ± 0.1	75.71 ± 0.43	64.69 ± 0.1	70.68 ± 0.41

We observe from this experiment that FDPIE reduces significantly the number of support vectors even when λ takes a small value (more than 84 % of support vectors are reduced when λ is set to 0.1). The results obtained from FDPIE are better than these of FSVM-L1 (71 % of support vectors are reduced). However, when λ is larger than 0.5, the number support vectors is dramatically reduced which negatively affects on the accuracy of classifiers. Therefore, to compare the sparsity in the solutions obtained from the l_0 and l_1 regularization with the same values of λ, in the experiment 2, we fixed the value of $\lambda = 0.5$ for both FDPIE and the concurrent method (the problem (18)). The results are reported in details in Table 2.

The experiment 2 shows that: with the same values of λ, FDPIE helps to reduce the number of support vectors (from more than 64 % to more than 90 %) while still ensures a good accuracy of classifiers. These results are better than these of FSVM-L1 which helps to reduce from more than 30 % to more than 73 % of the number of support vectors.

7 Conclusion

In this paper, we have developed an efficient approach based on DC programming and DCA for support vector reduction in Fuzzy Support Vector Machines. The l_0 regularization term is added to the dual form of FSVM model. Using the Concave function for approximation of zero-norm, we get a non-convex optimization problem. A DC decomposition and a DCA, namely FDPIE is presented to solve the resulting problem. Numerical results on several real datasets showed the robustness, effectiveness of the DCA based scheme. The effectiveness of the proposed algorithm as motivation for us to investigate other approximations of the zero norm and design corresponding DCAs for this important issue.

References

1. Bradley, P.S., Mangasarian, O.L.: Feature selection via concave minimization and support vector machines. In: Shavlik, J. (ed.) Machine Learning Proceedings of the Fifteenth International Conferences (ICML'98), pp. 82–90. Morgan Kaufmann, San Francisco (1998)
2. Brank, J., Grobelnik, M., Milic-Frayling, N., Mladenic, D.: Feature selection using linear support vector machines. In: Proceedings of the 3rd International Conference on Data Mining Methods and Databases for Engineering (2002)
3. Burges, C.J.: Simplified support vector decision rules. In: Machine learning-International Workshop then Conference, pp. 71–77 (1996)
4. Downs, T., Gates, K.E., Masters, A.: Exact simplification of support vector solutions. J. Mach. Learn. Res. **2**, 293–297 (2002)
5. Fan, J., Li, R.: Variable selection via nonconcave penalized likelihood and its Oracle Properties. J. Am. Stat. Assoc. **96**, 1348–1360 (2001)
6. Figueiredo, M.A.: Adaptive sparseness for supervised learning. IEEE Trans. Pattern Anal. Mach. Intell. **25**, 1150–1159 (2003)
7. Guyon, I., Elisseeff, A.: An introduction to variable and feature selection. J. Mach. Learn. Res. **3**, 1157–1182 (2003)
8. Hui, Z.: The adaptive lasso and its oracle properties. J. Am. Stat. Assoc. **101**(476), 1418–1429 (2006)
9. Huang, J., Ma, S., Zhang, C.H.: Adaptive lasso for sparse high-dimensional regression models. Stat. Sinica **18**, 1603–1618 (2008)
10. Huang, K., Zheng, D., Sun, J., Hotta, Y., Fujimoto, K., Naoi, S.: Sparse learning for support vector classification. Pattern Recogn. Lett. **31**, 1944–1951 (2010)
11. Huang, K., King, I., Lyu, M.R.: Direct zero-norm Optimization for feature selection. In: Eighth IEEE International Conference on Data Mining, pp. 840–850 (2008)
12. Le Thi, H.A., Pham Dinh, T.: The DC (Difference of convex functions) programming and DCA revisited with DC models of real world nonconvex optimization problems. Ann. Oper. Res. **133**, 23–46 (2005)
13. Le Thi, H.A., Belghiti, T., Pham Dinh, T.: A new efficient algorithm based on DC programming and DCA for Clustering. J. Global Optim. **37**, 593–608 (2006)
14. Le Thi, H.A., Le Hoai, M., Pham Dinh, T.: Optimization based DC programming and DCA for Hierarchical Clustering. Eur. J. Oper. Res. **183**, 1067–1085 (2007)
15. Le Thi, H.A., Le Hoai, M., Nguyen, V.V., Pham Dinh, T.: A DC Programming approach for feature selection in support vector machines learning. J. Adv. Data Anal. Classif. **2**(3), 259–278 (2008)
16. Le Thi, H.A., Nguyen, V.V., Ouchani, S.: Gene selection for cancer classification using DCA. In: Tang, C., Ling, C.X., Zhou, X., Cercone, N.J., Li, X. (eds.) ADMA 2008. LNCS (LNAI), vol. 5139, pp. 62–72. Springer, Heidelberg (2008)
17. Le Thi, H.A.: A new approximation for the ℓ_0-norm. Research Report LITA EA 3097, University of Lorraine (2012)
18. Le Thi, H.A., Pham Dinh, T., Le Hoai, M., Vo, X.T.: DC approximation approaches for sparse optimization. Eur. J. Oper. Res. **244**(1), 26–46 (2015)
19. Le Hoai, M., Le Thi, H.A., Nguyen, M.C.: Sparse semi-supervised support vector machines by DC programming and DCA. Neurocomputing **153**, 62–76 (2015)
20. Le Thi, H.A., Nguyen, M.C.: Efficient algorithms for feature selection in multi-class support vector machine. In: Advanced Computational Methods for Knowledge Engineering, Studies in Computational Intelligence, vol. 479, pp. 41–52, Springer (2013)
21. Le Thi, H.A., Huynh, V.N., Pham Dinh, T.: Exact penalty and error bounds in DC programming. J. Global Optim. Dedic. Reiner Horst. ISSN 0925-5001. doi:10.1007/s10898-011-9765-3 (2011)
22. Lee, J.Y., Mangasarian, O.L.: RSVM: reduced support vector machines. In: Proceedings of the First SIAM International Conference on Data Mining, pp. 5–7 (2001)

23. Lee, Y.J., Huang, S.Y.: Reduced support vector machines: a statistical theory. IEEE Trans. Neural Networks **18**, 1–13 (2007)
24. Lee, G.H., Taur, J.S., Tao, C.W.: A Robust Fuzzy support vector machine for two-class pattern classification. Int. J. Fuzzy Syst. **8**(2), 76–86 (2006)
25. Li, Q., Jiao, L., Hao, Y.: Adaptive simplification of solution for support vector machine. Pattern Recogn. **40**, 972–980 (2007)
26. Li, Y., Zhang, W., Lin, C.: Simplify support vector machines by iterative learning. Neural Inf. Process. Lett. Rev. **10**, 11–17 (2006)
27. Lin, C.F., Wang, S.D.: Fuzzy support vector machines. IEEE Trans. Neural Networks **13**(2), 464–471 (2002). doi:10.1109/72.991432
28. Jian, L., Xia, Z.: Fuzzy support vector machines based algorithm for peptide identification from Tandem mass spectra. Int. J. Pure Appl. Math. **76**(3), 439–447 (2012)
29. Nguyen, D., Ho, T.: An efficient method for simplifying support vector machines. In: Proceedings of the 22nd International Conference on Machine Learning, pp. 617–624 (2005)
30. Ong, C.S., Le Thi, H.A.: Learning sparse classifiers with difference of Convex functions Algorithms. Optim. Methods Softw. **28**, 4 (2013)
31. Dinh, P., T., Le Thi, H.A.: Convex analysis approach to d.c. programming: theory, algorithm and applications. Acta Mathematica Vietnamica **22**, 289–355 (1997)
32. Pham Dinh, T., Le Thi, H.A.: Optimization algorithms for solving the trust region subproblem. SIAM J. Optim. **2**, 476–505 (1998)
33. Pham Dinh, T., Le Thi, H.A.: Recent advances on DC programming and DCA. Trans. Comput. Intell. XIII Lect. Notes Comput. Sci. **8342**, 1–37 (2014)
34. He, Q., Congxin, W.: Membership evaluation and feature selection for fuzzy support vector machine based on fuzzy rough sets. Soft Comput. **15**, 1105–1114 (2011). doi:10.1007/s00500-010-0577-z
35. Rakotomamonjy, A.: Variable selection using SVM-based criteria. J. Mach. Learn. Res. **3**, 1357–1370 (2003)
36. Keerthi, S.S., Chapelle, O., DeCoste, D.: Building support vector machines with reduced classifier complexity. J. Mach. Learn. Res. **7**, 1493–1515 (2006)
37. Scholkopf, B., Mika, S., Burges, C.J., Knirsch, P., Muller, K.R., Ratsch, G.: Input space versus feature space in kernel-based methods. IEEE Trans. Neural Networks **10**, 1000–1017 (1999)
38. Shifei, D.I.N.G., Yaxiang, G.U.: A fuzzy support vector machine algorithm with dual membership based on hypersphere. J. Comput. Inf. Syst. **7**(6), 2028–2034 (2011)

DC Programming and DCA for Transmit Beamforming and Power Allocation in Multicasting Relay Network

Hoai An Le Thi, Thi Thuy Tran, Tao Pham Dinh and Alain Gély

Abstract This paper concerns a single-group multicasting relay network comprising of multiple amplify-and-forward relays forwarding signal from a single source to multiple destinations. The source, relays and destinations are equipped by single antennas. In this scenario, we deal with the problem of maximizing the minimum Quality of Service (QoS) assessed by the Signal to Noise Ratio (SNR) of the destinations subject to power constraints. This problem is actually a nonconvex optimization one with nonconvex constraints, which can be reformulated as a DC (Difference of Convex functions) program with DC constraints. We will apply an extension of DC programming and DCA (DC Algorithms) for solving it. Numerical experiments are carried out on several networks simulated under realistic conditions.

Keywords DC programming · DCA · Transmit beamforming · Power allocation · Multicasting relay network

1 Introduction and Related Works

To meet the higher and higher demand of wireless network users for quality of service, the next-generation wireless networks should be developed in the direction of offering more new techniques with the aim of achieving a better data rate compared

H.A. Le Thi · T.T. Tran (✉) · A. Gély
Laboratory of Theoretical and Applied Computer Science (LITA), UFR MIM,
University of Lorraine, Ile du Saulcy, 57045 Metz, France
e-mail: thi-thuy.tran@univ-lorraine.fr; thuytt@fpt.edu.vn

H.A. Le Thi
e-mail: hoai-an.le-thi@univ-lorraine.fr

A. Gély
e-mail: alain.gely@univ-lorraine.fr

T. Pham Dinh
Laboratory of Mathematics, INSA-Rouen University of Normandie, 76801
Saint-Etienne-du-Rouvray Cedex, France
e-mail: pham@insa-rouen.fr

© Springer International Publishing Switzerland 2016
T.B. Nguyen et al. (eds.), *Advanced Computational Methods
for Knowledge Engineering*, Advances in Intelligent Systems
and Computing 453, DOI 10.1007/978-3-319-38884-7_3

to the currently deployed networks. Nevertheless, the difficulty of obtaining a good data rate is often caused by the interference that arises due to the increased temporal and spectral reuse of resources. As a result, the novel techniques that exploit the spatial domain will contribute significantly in the efficient operation of future networks. Among them, multiple-input multiple-output (MIMO) antenna configurations, cooperative relays, and beamforming (BF) have been very active research fields in recent years, as these techniques enable interference mitigation.

In general, the literature on cooperative relaying has witnessed a remarkable increase in contributions recently. This technique plays an important role in improving three critical parameters of wireless networks. With multihop transmission permission, transmitters are brought closer to the receiver, thus reducing the path loss attenuation of the signal. Moreover, shadowing can be restricted by placing relay nodes in the positions of the obstacles that affect single-hop communications. In addition, multipath fading is alleviated due to the provision of independent propagation paths. In short, through cooperative relaying, coverage is extended, reliability is risen, and diversity can be harvested.

Beamforming is actually a signal processing technique in which BF matrices are used at transmitters and receivers for directional signal transmission and reception. The entries of BF matrices are chosen in such a way to satisfy a particular objective, such as the mean square error (MSE) or the signal-to-noise ratio (SNR). The beamforming technique is regarded as a powerful approach to receive, transmit, or relay signal-of-interests in a spatially selective way in the presence of interference and noise. Many recent works have used this technique to enhance coverage and data rate performance in amplify-and-forward (AF) relay networks. It has been applied to various relay network architectures, ranging from single-user networks to peer-to-peer networks and then to multi-user multicasting networks. This technique can be classified into two categories: distributed beamforming through using nonconnected relays [1, 15] and centralized beamforming through using a connected antenna array [6, 7].

To address the beamforming problem, which is in essence a nonconvex quadratically constrained quadratic optimization problem (QCCP), computationally efficient various approaches in direction of approximating the feasible set have been applied [14, 19, 20]. These methods tend to narrow the feasible set that may make the resulting problem infeasible. To overcome this difficulty, the technique of searching the feasible initial point by introducing a slack variable into the approximated constraints was mentioned in [3]. Another method, which should be considered, is the combination of penalty technique and method of introducing a slack variable as mentioned above [9].

Another technique to solve the QCCP is the outer approximation which reformulates the QCQP by a convex semidefinite program (SDP) that is tractable after eliminating the rank constraints. However, this semidefinite relaxation (SDR) technique makes the feasible set larger, thus it only provides an upper/lower bound for objective value of maximization/minimization beamforming problem. In case the SDR solution does not belong to the original feasible set, randomization techniques have been deployed to generate feasible points that are in general suboptimal [17, 20].

In this paper, we take account of a distributed beamforming scheme for single-group multicasting using a network of amplify-and-forward (AF) relays. Our objective is that we not only design the BF vectors to form the beam to intended destination, but also find the scaling factors to control power between difference time slots and between the source and the relays, with the purpose of maximizing the minimum SNR subject to power constraints. The max-min fairness is the well-known criterion and widely used in many works such as [2, 4, 18]. The model used in this paper was introduced in [16] and was solved based on both SDR technique and concave-convex procedure (CCCP) that is in fact a DCA based algorithm. In this paper, the approach based on DC programming and DCA with a novel DC decomposition in combination with the penalty technique is proposed to address this max-min fairness optimization problem.

2   Transmit Beamforming and Power Allocation in Multicasting Relay Networks

In this section, we briefly restate the problem formulated in [16].

Consider a wireless system comprising of a single source, R relays and M destinations. The source, each relay and each destination are equipped by single antenna. In this model, two data symbols are simultaneously processed in a four time slot scheme. All channels in the network are supposed to be frequency flat and constant over the considered four time slots.

In the first and second time slot, the source transmits the data symbols s_1 and s_2^* to the relays and the destinations respectively. Both symbols are multiplied by the same coefficient $p_1 \in \mathbb{R}$ before being sending. The relays, in the first and second time slot, receive the following signals

$$r_1 = fp_1 s_1 + n_{R,1}, \ r_2 = fp_1 s_2^* + n_{R,2}, \tag{1}$$

where $n_{R,1} \in \mathbb{R}^R$ and $n_{R,2} \in \mathbb{R}^R$ are the relay noise vectors of the first and the second time slot, respectively, and $f \in \mathbb{R}^R$ is the vector of the channel coefficients between the source and the relays. The signal $d_{m,1}$ and $d_{m,2}$ received by the mth destination in the first and second time slot are respectively computed by

$$d_{m,1} = h_m p_1 s_1 + n_{D,m,1}, \ d_{m,2} = h_m p_1 s_2^* + n_{D,m,2}, \tag{2}$$

where h_m is the channel coefficient from the source to the mth destination and $n_{D,m,1}, n_{D,m,2}$ are the noise at the mth destination in the first and second time slot, respectively.

$$t_3 = W_1 r_1 + W_2 r_2^*, \ t_4 = -W_2 r_1^* + W_1 r_2, \tag{3}$$

where $W_1 \triangleq \mathrm{diag}(w_1^H)$, $W_2 \triangleq \mathrm{diag}(w_2^H)$, and $w_1 = \left[w_{1,1}, ..., w_{R,1}\right]^T$,

$w_2 = \left[w_{1,2}, ..., w_{R,2}\right]^T$ are the complex $R \times 1$ beamforming vectors. At the same time, the source sends the signals $p_3 s_1 + p_4 s_2$ and $-p_4 s_1^* + p_3 s_2^*$, respectively to the destinations, where p_3, p_4 are complex weights. The signals received by the mth destination in the third and fourth time slot are calculated by

$$
\begin{aligned}
d_{m,3} &= g_m^T t_3 + h_m(p_3 s_1 + p_4 s_2) + n_{D,m,3}, \\
d_{m,4} &= g_m^T t_4 + h_m(-p_4 s_1^* + p_3 s_2^*) + n_{D,m,4},
\end{aligned}
\tag{4}
$$

where $g_m \in \mathbb{R}^R$ is the vector of the complex channel coefficients between the relays and the mth destination and $n_{D,m,3}, n_{D,m,4}$ are the receiver noise at the mth destination in the third and fourth time slot, respectively. It is assumed that the noise processes in the network are spatially and temporally independent and complex Gaussian distributed. The noise power at the destinations equals to $E\{|n_{D,m,q}|^2\} = \sigma_D^2$, $q \in \{1, 2, 3, 4\}$, and the noise at the relays has distribution $n_{R,1} \sim \mathcal{N}(0_R, \sigma_R^2 I_R), n_{R,2} \sim \mathcal{N}(0_R, \sigma_R^2 I_R)$. Let us denote the vector of the received signals at the mth destination by $d_m \triangleq \left[d_{m,1}, d_{m,2}^*, d_{m,3}, d_{m,4}^*, \right]^T$, the vector of the noise at the mth destination by n_m, the equivalent channel matrix by Z_m and using the Eqs. (1), (3) and (2), the received signals of the four time slots can be jointly written as

$$
d_m = Z_m s + n_m
\tag{5}
$$

where

$$
s = [s_1, s_2]^T, n_m \triangleq
\begin{bmatrix}
n_{D,m,1} \\
n_{D,m,2}^* \\
w_1^H G_m n_{R,1} + w_2^H G_m n_{R,2}^* + n_{D,m,3} \\
-w_2^T G_m^H n_{R,1} + w_1^T G_m^H n_{R,2}^* + n_{D,m,4}^*
\end{bmatrix},
H_m \triangleq
\begin{bmatrix}
p_1 h_m & 0 \\
0 & (p_1 h_m)^* \\
z_{m,1} & z_{m,2} \\
-z_{m,2}^* & z_{m,1}^*
\end{bmatrix}
\tag{6}
$$

with $z_{m,1} \triangleq p_1 w_1^H G_m f + p_3 h_m$, $z_{m,2} \triangleq p_1 w_2^H G_m f^* + p_4 h_m$, $G_m \triangleq \mathrm{diag}(g_m)$. It is easy to verify that $E(n_m n_m^H) = \mathrm{blkdiag}([\sigma_D^2 I_2, \sigma_{m,34}^2 I_2])$, where

$$
\sigma_{m,34}^2 \triangleq \sigma_R^2 (w_1^H \mathcal{G}_m w_1 + w_2^H \mathcal{G}_m w_2) + \sigma_D^2, \quad \mathcal{G}_m \triangleq G_m G_m^H.
$$

The signal-to-noise ratio for both data symbols at the mth destination is shown in [16] by the formula below

$$
\mathrm{SNR}_m = \frac{p_1^2 |h_m|^2}{\sigma_D^2} + \frac{|z_{m,1}|^2 + |z_{m,2}|^2}{\sigma_{m,34}^2}.
\tag{7}
$$

By denoting $p = \frac{1}{p_1^2}$, $w = [w_1^T, p_3^*/p1, w_2^T, p_4^*/p_1]^T$, $B_m = \mathrm{blkdiag}([\sigma_R^2 \mathcal{G}_m, 0, \sigma_R^2 \mathcal{G}_m, 0])$,

$$A_m = \text{blkdiag}([a_{m,1}a_{m,1}^H, \ a_{m,2}a_{m,2}^H]), \text{ with } a_{m,1} = \begin{bmatrix} G_m f \\ h_m \end{bmatrix}, a_{m,2} = \begin{bmatrix} G_m f^* \\ h_m \end{bmatrix} \text{ and from}$$

the formula of $z_{m,1}, z_{m,2}$, (7) can be rewritten in the following form

$$\text{SNR}_m(w,p) = \frac{w^H A_m w}{(w^H B_m w + \sigma_D^2)p} + \frac{|h_m|^2}{\sigma_D^2 p}. \tag{8}$$

The optimization problem considered in [16] is of the following form

$$\max_{w,p} \ \min_{m \in \{1,..,M\}} \ \text{SNR}_m(w,p) \tag{9}$$

$$\text{s.t.} \quad (w,p) \in \Omega,$$

where $\Omega = \{(w,p)\}$ satisfies the constraints below

positivity: $p > 0$

individual relay power: $p_r(w,p) = w^H D_r w/p + w^H E_r w \le p_{r,max} \ \forall r \in \{1, ..., R\}$,

relay sum power: $\displaystyle\sum_{r=1}^{R} p_r(w,p) = \sum_{r=1}^{R} (w^H D_r w/p + w^H E_r w) \le P_{R,max}$,

source power: $P_S(w,p) = 2/p + w^H S_r w/p \le P_{S,max}$,

total power: $P_T(w,p) = 2/p + w^H S_r w/p + 2 \displaystyle\sum_{r=1}^{R} (w^H D_r w/p + w^H E_r w) \le P_{T,max}$,

where $D_r \triangleq \text{blkdiag}\left(\left[\hat{D}_r, \hat{D}_r\right]\right)$, $E_r \triangleq \text{blkdiag}\left(\left[\hat{E}_r, \hat{E}_r\right]\right)$ and $S \triangleq \text{blkdiag}\left(\left[\hat{S}, \hat{S}\right]\right)$, in which $\hat{D}_r$ is a $(R+1) \times (R+1)$ matrix with all entries equal to zero except (r,r)-entry equal to $|f_r|^2$, $\hat{E}_r$ is a $(R+1) \times (R+1)$ matrix having σ_R^2 as its rth diagonal entry and zeros elsewhere, $\hat{S}$ is a $(R+1) \times (R+1)$ diagonal matrix with $(R+1, R+1)$-entry equal to 2 and zeros elsewhere.

In multicast networks, a trade-off between the transmitted power and the QoS at the intended receivers has to be met. The best trade-off is achieved by solving an optimization problem, where the beamforming weight vectors and power scaling factors are the optimization variables. In this model, the objective is to find the beamforming weight vectors as well as power scaling factors (i.e. (w,p)) to maximize the minimum QoS measured in terms of the SNR at the destinations subject to power constraints. The worst SNR is an important limiting value in multicasting application because it determines the common information rate. Maximizing the worst SNR is to ensure fairness among users, avoid the existence of users with very poor SNR. This max-min fairness criterion has been used in many previous works as mentioned before.

Note that the quadratic form $w^H A w$ and the fraction of the quadratic form $w^H A w$ and the linear term p are convex provided that A is a Hermitian and semidefinite matrix. Therefore the constraint set Ω mentioned above is convex.

Denote $1/t = \min_{m \in \{1,...,M\}} \mathrm{SNR}_m$, then the problem (9) can be equivalently reformulated as

$$\min_{w,p,t} \quad t \tag{10}$$

$$\text{s.t.} \quad \mathrm{SNR}_m(w,p) \geq 1/t \ \forall m \in \{1,...,M\}, \tag{11}$$

$$(w,p) \in \Omega, \ t > 0.$$

This problem is nonconvex with the nonconvex constraints (11) that can be reformulated as a difference of two convex functions (called DC functions). Therefore, a DC Programming and DCA based approach, which is shown by many works as an efficient method for coping with the difficulty caused by the nonconvexity in optimization problems, is an appropriate choice.

3 Solution Methods by DC Programming and DCA

3.1 A Brief Overview of DC Programming and DCA

DC Programming and DCA constitute the backbone of smooth/nonsmooth nonconvex programming and global optimization. They were introduced by Pham Dinh Tao in 1985 in their preliminary form and have been extensively developed by Le Thi Hoai An and Pham Dinh Tao since 1994 to become now classic and more and more popular. DCA is a continuous primal dual subgradient approach. It is based on local optimality and duality in DC programming in order to solve standard DC programs, which are of the form

$$\alpha = \inf \{ f(x) := g(x) - h(x) \ : \ x \in \mathbb{R}^n \}, \quad (P_{dc})$$

with $g, h \in \Gamma_0(\mathbb{R}^n)$, which is a set of lower semi-continuous proper convex functions on $\mathbb{R}^n$. Such a function f is called a DC function, and $g - h$, a DC decomposition of f, while the convex functions g and h are DC components of f. A constrained DC program whose feasible set C is convex always can be transformed into an unconstrained DC program by adding the indicator function of C to the first DC component.

Recall that, for a convex function ϕ, the subgradient of ϕ at x_0, denoted as $\partial \phi(x_0)$, is defined by $\partial \phi(x_0) := \{ y \in \mathbb{R}^n \ : \ \phi(x) \geq \phi(x_0) + \langle x - x_0, y \rangle, \forall x \in \mathbb{R}^n \}$.

The main principle of DCA is quite simple, that is, at each iteration of DCA, the convex function h is approximated by its affine minorant at $y^k \in \partial h(x^k)$, and it leads to solving the resulting convex program.

$$x^{k+1} \in \arg \min_{x \in \mathbb{R}^n}\{g(x) - h(x^k) - \langle x - x^k, y^k \rangle\}, \text{with } y^k \in \partial h(x^k). \quad (P_k)$$

The computation of DCA is only dependent on DC components g and h but not the function f itself. Actually, there exist infinitely many DC decompositions corresponding to each DC function and they generate various versions of DCA. Choosing an appropriate DC decomposition plays a key role since it influences on the properties of DCA such as convergence speed, robustness, efficiency, globality of computed solutions,...DCA is thus a philosophy rather than an algorithm. For each problem we can design a family of DCA based algorithms. To the best of our knowledge, DCA is actually one of the rare algorithms for nonsmooth nonconvex programming which allow to solve large-scale DC programs. DCA was successfully applied for solving various nonconvex optimization problems, which quite often gave global solutions and is proved to be more robust and more efficient than related standard methods [11–13] and the list of reference in [8]. The convergence properties of DCA and its theoretical basis is analyzed and proved completely in [10–12].

Recently, the extension of DC programming and DCA was studied in [9] to solve general DC programs with DC constraints as follows

$$\min_x \ f_0(x) \tag{12}$$
$$\text{s.t} \ f_i(x) \le 0 \ \forall i = 1, ..., m,$$
$$x \in C,$$

where $C \subseteq \mathbb{R}^n$ is a nonempty closed convex set; $f, f_i : \mathbb{R}^n \to \mathbb{R}(i = 0, 1, ..., m)$ are DC functions. It is apparent that this class of nonconvex programs is the most general in DC programming and as consequence it is more challenging to deal with than standard DC programs. Two approaches for general DC programs were proposed in [9] to overcome the difficulty caused by the nonconvexity of the constraints. Both approaches are built on the main idea of the philosophy of DC programming and DCA, that is approximating (12) by a sequence of convex programs. The former was based on penalty techniques in DC programming while the latter was relied on the convex inner approximation method. Because we use the second approach to solve the problems mentioned in this article, we presented herein its main scheme.

Since $f_i(i = 0, ..., m)$ are DC functions, they can be decomposed into the difference of two convex functions $f_i(x) = g_i(x) - h_i(x)$, $x \in \mathbb{R}^n, i = 0, ..., m$. By linearizing the concave part of DC decompositions of all DC objective function and DC constraints, we derive sequential convex subproblems of the following form:

$$\min_x \ g_0(x) - \langle y_0^k, x \rangle \tag{13}$$
$$\text{s.t} \ g_i(x) - h_i(x^k) - \langle y_i^k, x - x^k \rangle \le 0 \ \forall i = 1, ..., m,$$
$$x \in C,$$

where $x^k \in \mathbb{R}^n$ is a point at the current iteration , $y_i^k \in \partial h_i(x^k) \ \forall i = 0, ..., m$.

This linearization introduces an inner convex approximation of the feasible set of (12). However, it may lead to infeasibility of convex subproblems (13). The relaxation technique was proposed to confront this difficulty. Instead of (13), we consider the relaxed subproblem:

$$\min_{x} \ g_0(x) - <y_0^k, x> +\beta_k t \tag{14}$$

$$\text{s.t } g_i(x) - h_i(x^k) - <y_i^k, x - x^k> \ \le t \ \forall i = 1, ..., m,$$

$$x \in C, t \ge 0,$$

where β_k is a penalty parameter. It is easy to realize that the relaxed subproblem (14) is always feasible. Furthermore, the Slater constraint qualification is satisfied for the constraints of (14), thus the Karush-Kuhn-Tucker (KKT) optimality condition holds for some solutions x^{k+1}, t^{k+1}. Thus, there exits some $\lambda_i^{k+1} \in \mathbb{R}, i = 1, .., m$ and $\mu^{k+1} \in \mathbb{R}$ such that

- $0 \in \partial g_0(x^{k+1}) - y_0^k + \sum_{i=0}^{m} \lambda_i^{k+1}(\partial g_i(x^{k+1}) - y_i^k) + N(C, x^{k+1}),$

- $\beta_k - \sum_{i=1}^{m} \lambda_i^{k+1} - \mu^{k+1} = 0,$

- $g_i(x^{k+1}) - h_i(x^k) - <y_i^k, x^{k+1} - x^k> \le t^{k+1}, \lambda_i^{k+1} \ge 0 \ \forall i = 1, ..., m, x^{k+1} \in C,$

- $\lambda_i^{k+1}(g_i(x^{k+1}) - h_i(x^k) - <y_i^k, x^{k+1} - x^k> -t^{k+1}) = 0, \forall i = 1, ..., m,$

- $t^{k+1} \ge 0, \mu^{k+1} \ge 0, t^{k+1}\mu^{k+1} = 0.$

The DCA scheme for the general DC program (12) is proposed as follows:

- **Initialization**. Choose an initial point $x^0; \delta_1, \delta_2 > 0$, an initial penalty parameter $\beta_1 > 0$. Set $0 \longleftarrow k$.
- **Repeat**.
  Step 1. Calculating $y_i^k \in \partial h_i(x^k), \ i = 0, .., m,$
  Step 2. Compute (x^{k+1}, t^{k+1}) as the solution of (14), and the associated Lagrange multipliers $(\lambda^{k+1}, \mu^{k+1}),$
  Step 3. Penalty parameter update.
  compute $r_k = \min\{\|x^{k+1} - x^k\|^{-1}, \|\lambda^{k+1}\|_1 + \delta_1\}$
  and set $\beta_{k+1} = \begin{cases} \beta_k & \text{if } \beta_k \ge r_k, \\ \beta_k + \delta_2 & \text{if } \beta_k < r_k. \end{cases}$
  Step 4. $k \leftarrow k + 1,$
- **Until** $x^{k+1} = x^k$ and $t^{k+1} = 0$.

The proof of global convergence of the above algorithm is shown in Theorem 2 of the article [9].

3.2 DC Programming and DCA for the Problem (10)

DC decomposition for nonconvex constraints (11)

We have

$$\text{SNR}_m \geq \frac{1}{t} \iff \frac{w^H A_m w}{(w^H B_m w + \sigma_D^2)p} + \frac{|h_m|^2}{\sigma_D^2 p} \geq \frac{1}{t}$$
$$\iff G_m(w,p,t) - H_m(w,p,t) \leq 0,$$

where

$$G_m(w,p,t) = \left(1 + \frac{|h_m|^2}{\sigma_D^2}\right)t^2 + (w^H A_m w)^2 + (p + w^H B_m w + \sigma_D^2)^2 + \frac{|h_m|^2}{\sigma_D^2}(w^H B_m w + \sigma_D^2)^2$$

and

$$H_m(w,p,t) = (t + w^H A_m w)^2 + \frac{|h_m|^2}{\sigma_D^2}(t + w^H B_m w + \sigma_D^2)^2 + p^2 + (w^H B_m w + \sigma_D^2)^2).$$

The convexity of G_m and H_m are deduced from that of the quadratic form $w^H A w$ with some Hermitian, semidefinite matrices A and that of its composition function.

A linear minorant of $H_m(w,p,t)$ at (w^k, p^k, t^k) is given by

$$\overline{H}_m(w^k, p^k, t^k) = M_k^2 + V N_k^2 + (p^k)^2 + C_k^2 + 2(M_k + V N_k)(t - t^k) + 2p^k(p - p^k)$$
$$+ Re\{(w - w^k)^H T_k w^k\},$$

where $M_k = t^k + (w^k)^H A_m w^k$, $C_k = \sigma_D^2 + (w_k)^H B_m w^k$, $N_k = t^k + C_k$, $V = \frac{|h_m|^2}{\sigma_D^2}$, $T_k = 4M_k A_m + 4(V N_k + C_k)B_m$.

The main idea of the DCA is that, at each iteration, we replace the second convex function of the DC function by its linear minorant, which raises sequential convex subproblems of the following form

$$\min_{w,p,t} \quad t \tag{15}$$

$$\text{s.t.} \quad G_m(w,p,t) - \overline{H}_m(w^k, p^k, t^k) \leq 0 \ \forall m \in \{1,...,M\}, \tag{16}$$
$$(w,p) \in \Omega, \ t > 0.$$

However, this linearization introduces an inner convex approximation of the feasible set of (10) since $G_m(w, p, t) - H_m(w, p, t) \leq G_m(w, p, t) - \overline{H}_m(w^k, p^k, t^k)$, which can lead to infeasibility of the convex subproblems. Thus, we propose a relaxation technique to deal with the feasibility of subproblem. More specifically, we introduce a slack variable for the constraints (16) and penalize it into the objective function to prevent it from becoming large. As a consequence, instead of solving the convex subproblem (15), we address the relaxed one below

$$\min_{w,p,t,s} \quad t + \tau s \tag{17}$$

$$\text{s.t.} \quad G_m(w, p, t) - \overline{H}_m(w^k, p^k, t^k) \leq s \quad \forall m \in \{1, ..., M\},$$

$$(w, p) \in \Omega, \ t > 0,$$

where τ is a penalty parameter.

DCA scheme for DC program (10)

- **Initialization.** Choose an initial point $x^0 = (w^0, p^0, t^0)$, a penalty parameter $\tau = \tau_0$, and two tolerances ϵ_1, ϵ_2. $0 \longleftarrow k$.

- **Repeat.**
  Step 1. For each $k, x^k = (w^k, p^k, t^k)$ is known, solving the relaxed convex subproblem (17) to find $x^{k+1} = (w^{k+1}, p^{k+1}, t^{k+1})$.
  Step 2. $k \leftarrow k + 1$.

- **Until** either $\frac{t^k - t^{k+1}}{t^k + 1} < \epsilon_1$ or $\frac{\|x^k - x^{k+1}\|}{\|x^k\| + 1} < \epsilon_2$ and $|s| < \epsilon_2$.

---

4 Experimental Results

In this section, we give the numerical performance obtained by DCA and then compare them with those obtained by Max-Min-CCCP and SDR2D based algorithms, which are mentioned in [16]. In all experiments, the average minimum achieved rate, which is calculated by $\frac{1}{2} \log_2(1 + \min_{m=1,...,M}\{\text{SNR}_m\})$, is used for comparing among algorithms.

4.1 Simulated Datasets and Parameter Setting

In our experiments, we consider a network with $R = 10$ relay nodes. The channel coefficients are assumed to be independent from each other. Specifically, $f_i, d_m, g_{m,i}$ are modeled as

$$f_i = \bar{f}_i + \hat{f}_i \; \forall i \in \{1,..,R\},$$
$$d_m = \bar{d}_m + \hat{d}_m \; \forall m \in \{1,..,M\},$$
$$g_{m,i} = \bar{g}_{m,i} + \hat{g}_{m,i} \; \forall m \in \{1,..,M\}, \; \forall i \in \{1,..,R\},$$

where $\bar{f}_i, \bar{d}_m, \bar{g}_{m,i}$ are complex channel mean and $\hat{f}_i, \hat{d}_m, \hat{g}_{m,i}$ are zero-mean random variables $\forall m \in \{1,..,M\}, \; \forall i \in \{1,..,R\}$. According to [5], the channel mean $\bar{f}_i, \bar{d}_m, \bar{g}_{m,i}$ can be modeled, respectively, as

$$\bar{f}_i = \frac{exp(\sqrt{-1}\Theta_i)}{\sqrt{\Gamma_f}}, \bar{d}_m = \frac{exp(\sqrt{-1}\Omega_i)}{\sqrt{\Gamma_d}}, \bar{g}_{m,i} = \frac{exp(\sqrt{-1}Y_{m,i})}{\sqrt{\Gamma_g}},$$

where the random angles $\Theta_i, \Omega_i, Y_{m,i}$ are chosen to be uniformly distributed on the interval $[0, 2\pi] \; \forall m \in \{1,..,M\}, \; \forall i \in \{1,..,R\}$, and $\Gamma_f, \Gamma_d, \Gamma_g$ are positive constants, which indicate the uncertainty in the channel coefficients. Moreover, the variances of the random variables are given by

$$E\{|\hat{f}_i|^2\} = \frac{\Gamma_f}{\Gamma_f + 1}, E\{|\hat{d}_m|^2\} = \frac{\Gamma_d}{\Gamma_d + 1}, E\{|\hat{g}_{m,i}|^2\} = \frac{\Gamma_g}{\Gamma_g + 1}.$$

In this paper, we choose $\Gamma_f = \Gamma_d = \Gamma_g = -10$ dB. The noise powers at the relays and the destinations are set to $\sigma_R^2 = \sigma_D^2 = 1$. The maximum transmit power values are chosen such that $P_{S,max} = P_{T,max}/2, P_{R,max} = P_{T,max}/3$ and $p_{r,max} = P_{T,max}/15$. The total transmit power value, $P_{T,max}$, and the number of destinations, M, in the network are set differently in various experiments. More particularly, $P_{T,max}$ is chosen from the set $\{5, 10, 15, 20\}$ and M is chosen from the set $\{20, 40, 60, 80, 100\}$. The tolerances in DCA scheme are set to $\epsilon_1 = 10^{-3}, \; \epsilon_2 = 10^{-8}$.

4.2 Numerical Results and Comments

Table 1 describes the average minimum rate versus the number of destinations M in case of $P_{T,max} = 5$ dBm. It is observed from this table that the values of minimum rate obtained by DCA are better than those obtained by Max-Min CCCP. In other words, the minimum rate achieved by DCA is closer the upper bound of this value achieved by R2-SDR2D based algorithm. The gap between minimum rate obtained from DCA and R2-SDR2D based algorithm tends to be wider when M increases due to the fact that the rank of SDR solution matrix increases with the rise of M.

Table 2 depicts the average minimum rate versus $P_{T,max}$ in case of $M = 100$. It can be realized from this table that DCA obtains the better minimum rates that are closer to its upper bounds achieved by R2-SDR2D based algorithm, as compared to Max-Min-CCCP.

Table 1 Minimum rate versus number of destinations M

M	Max-Min-CCCP	DCA	R2-SDR2D
20	0.4709	0.4846	0.4854
40	0.3320	0.3600	0.3705
60	0.2733	0.2837	0.3078
80	0.2480	0.2653	0.2965
100	0.2392	0.2439	0.2633

Table 2 Minimum rate versus total power $P_{T,max}$

$P_{T,max}$	Max-Min-CCCP	DCA	R2-SDR2D
5	0.2392	0.2439	0.2633
10	0.4072	0.4242	0.4704
15	0.5391	0.5913	0.6139
20	0.7240	0.7388	0.7511

5   Conclusions

In this article, we propose a novel DC decomposition for DC constraints and use DCA in combination with relaxation technique and penalty method to solve the max-min fairness optimization problem. The results obtained by DCA are compared to those obtained by two algorithms R2-SDR2D and Max-Min-CCCP mentioned in [16]. The experimental performances show the efficiency of the proposed new DC decomposition as well as the combination of DCA with the techniques mentioned above. This approach can be applied to deal with numerous nonconvex optimization problems with DC constraints, which are commonly encountered in communication systems.

Acknowledgments This research is funded by Foundation for Science and Technology Development of Ton Duc Thang University (FOSTECT), website: http://fostect.tdt.edu.vn, under Grant FOSTECT.2015.BR.15.

References

1. Bornhorst, N., Pesavento, M., Gershman, A.B.: Distributed beamforming for multi-group multicasting relay networks. IEEE Trans. Signal Process. **60**(1), 221–232 (2011)
2. Chen, H., Greshman, B., Shahbazpanahi, S., Gazor, S.: Filter-and-forward distributed beamforming in relay networks with frequency selective fading. IEEE Trans. Signal Process. **58**, 1251–1262 (2010)
3. Cheng, Y., Pesavento, M.: Joint optimization of source power allocation and distributed relay beamforming in multiuser peer-to-peer relay networks. IEEE Trans. Signal Process. **60**(6), 2962–2973 (2012)

4. Dartmann, G., Zandi, E., Ascheid, G.: Equivalent quasi-convex form of the multicast max-min beamforming problem. IEEE Trans. Veh. Tech. **62**(9), 4643–4648 (2013)
5. Havary, V., Shahbazpanahi, S., Grami, A., Luo, Z.Q.: Distributed beamforming for relat networks based on second-order statistics of the channel state information. IEEE Trans. Signal Process. **56**(9), 4306–4316 (2008)
6. Kqripidis, E., Sidiropoulos, N.D., Luo, Z.Q.: Quality of service and max-min fair transmit beamforming to multiple co-channel multicast group. IEEE Trans. Signal Process. **56**, 1268–1279 (2008)
7. Law, K.L., Wen, X., Pesavento, M.: General-rank transit beamforming for multi-group multi-casting networks using OSTBC. In: Proceedings IEEE SPAWC'13, pp. 475–479. IEEE (June 2013)
8. Le Thi, H.A.: DC Programming and DCA. http://www.lita.univ-lorraine.fr/~lethi/
9. Le Thi, H.A., Huynh, V.N., Pham Dinh, T.: DC Programming and DCA for General DC Programs, vol. 282. Springer (2014)
10. Le Thi, H.A., Pham Dinh, T.: The DC (difference of convex functions) programming and DCA revisited with DC models of real world nonconvex optimization problems. Ann. Oper. Res. **133**, 23–46 (2005)
11. Pham Dinh, T., Le Thi, H.A.: Convex analysis approach to DC programming: theory, algorithms and applications. Acta Math. Vietnam. **22**(1), 289–357 (1997)
12. Pham Dinh, T., Le Thi, H.A.: Optimization algorithms for solving the trust region subproblem. SIAM J. Optim. **8**, 476–505 (1998)
13. Pham Dinh, T., Le Thi, H.A.: Recent Advances in DC Programming and DCA, vol. 8342. Springer, Berlin (2014)
14. Phan, K., Le-Ngoc, T., Vorobyov, S.A., Tellambura, C.: Power allocation in wireless multi-user relay networks. IEEE Trans. Wirel. Commun. **8**(5), 2535–2545 (2009)
15. Schad, A., Law, K.L., Pesavento, M.: A convex inner approximation techniqe for rank-two beamforming in multicasting relay networks. In: Proceedings Europen Signal Processing Conference, pp. 1369–1373 (August 2012)
16. Schad, A., Law, K., Pesavento, M.: Rank-two beamforming and power allocation in multicasting relay networks. IEEE Trans. Signal Process. **63**(13), 3435–3447 (2015)
17. Sidiropoulos, N.D., Davidson, T.N., Luo, Z.Q.: Transmit beamforming for physical-layer multicasting. IEEE Trans. Signal Process. **54**(6), 2239–2251 (2006)
18. Song, B., Lin, Y.H., Cruz, R.: Weighted max-min fair beamforming, power control and scheduling for a MISO downlink. IEEE Trans. Wirel. Commun. **7**, 464–469 (2008)
19. Wen, X., Law, K.L., Alabed, S.J., Pesavento, M.: Rank-two beamforming for single-group multicasting network using OSTBC. In: Proceedings IEEE SAM'012, pp. 69–72. IEEE, Hoboken, USA (2012)
20. Wu, S.X., Ma, W.K., So, A.M.: Physical layer multicasting by stochastic transmit beamforming and Alamouti space-time coding. IEEE Trans. Signal Process. **61**(17), 4230–4245 (2013)

Solving an Infinite-Horizon Discounted Markov Decision Process by DC Programming and DCA

Vinh Thanh Ho and Hoai An Le Thi

Abstract In this paper, we consider a decision problem modeled by Markov decision processes (written as MDPs). Solving a Markov decision problem amounts to searching for a policy, in a given set, which optimizes a performance criterion. In the considered MDP problem, we address the discounted criterion with the aim of characterizing the policies which provide the best sequence of rewards. In the literature, there are three main approaches applied to solve MDPs with a discounted criterion: linear programming, value iteration and policy iteration. In this paper, we are interested in the optimization approach to the discounted MDPs. Along this line, we describe an optimization model by studying the minimization of the different norms of Optimal Bellman Residual. In general, it can be formulated as a DC (Difference of Convex functions) program for which the unified DC programming and DCA (DC Algorithms) are applied. In our works, we propose a new optimization model and a suitable DC decomposition for the model of MDPs. Numerical experiments are performed on the stationary Garnet problems. The comparative results with the linear programming method for the discounted MDPs illustrate the efficiency of our proposed approach in terms of the quality of the obtained solutions.

Keywords Markov decision process · DC programming · DCA

1 Introduction

Markov decision processes (MDPs) are defined as controlled stochastic processes satisfying the Markov property and assigning reward values to state transitions [3, 29]. Many problems such as (stochastic) planning problems, reinforcement learning

---

The original version of this chapter was revised: The missing reference has been included.The correction to this chapter can be found at https://doi.org/10.1007/978-3-319-38884-7_21

V.T. Ho (✉) · H.A. Le Thi
Laboratory of Theoretical and Applied Computer Science EA 3097, University of Lorraine, Ile du Saulcy, 57045 Metz, France
e-mail: vinh-thanh.ho@univ-lorraine.fr

H.A. Le Thi
e-mail: hoai-an.le-thi@univ-lorraine.fr

© Springer International Publishing Switzerland 2016   43
T.B. Nguyen et al. (eds.), *Advanced Computational Methods for Knowledge Engineering*, Advances in Intelligent Systems and Computing 453, DOI 10.1007/978-3-319-38884-7_4

problems, learning robot control, game playing problems and other learning problems in stochastic domains have successfully been modeled in terms of an MDP [4–7, 10]. Applications of the MDP framework to real-life problems are growing everyday in various domains such as industrial processes management, agro-ecosystems, robotics, military operations management [4–7, 10].

In this paper, we address the MDP problems with a discounted, infinite-time horizon criterion. Formally, such a finite MDP can be described by the 5-tuple $< S, A, R, P, \gamma >$ where $S = \{s_i\}_{i=1,\ldots,N_s}$ is a finite state space of the process, $A = \{a_i\}_{i=1,\ldots,N_A}$ is a finite action space of the process, $R : S \times A \to \mathbb{R}$ is the reward function defined on state-action transitions, $P : S \times A \times S \to [0, 1]$ is the state transition probability function in which $P(s'|s, a)$ represents the probability of transition from $s \in S$ to $s' \in S$ upon taking action $a \in A$ and $\gamma \in (0, 1)$ is a discount factor. A stationary, deterministic policy π for the MDP is a mapping $\pi : S \to A$ where $\pi(s)$ is the action which the agent takes at the state s. In order to quantify the quantity of policy π, we define the value function V^π. For a given policy π, the value function $V^\pi \in \mathbb{R}^S$ is defined as

$$V^\pi(s) = \mathbb{E}^\pi \left[\sum_{i=0}^{\infty} \gamma^i R(s_i, \pi(s_i))|s_0 = s \right], \forall s \in S.$$

Moreover, the optimal policy $\pi^* \in S^A$ is defined as: for all $s \in S$,

$$\pi^*(s) \in \operatorname{argmax}_{\pi \in S^A} V^\pi(s).$$

A policy $\pi \in S^A$ is greedy with respect to a function $V \in \mathbb{R}^S$ if for all $s \in S$,

$$\pi(s) \in \operatorname{argmax}_{a \in A} \left(R(s, a) + \gamma \sum_{s' \in S} P(s'|s, a)V(s') \right).$$

The function $V^* \in \mathbb{R}^S$ defined as $V^* = V^{\pi^*}$ is said the optimal value function. As a result, the optimal policy π^* is greedy with respect to the optimal value function V^*. Specially, V^* is a unique fixed-point of the contracting Bellman optimal operator $T^* : \mathbb{R}^S \to \mathbb{R}^S$ defined as follows: for any real function V defined on S, we have

$$T^*V(s) = \max_{a \in A} \left\{ R(s, a) + \gamma \sum_{s' \in S} P(s'|s, a)V(s') \right\}, \forall s \in S. \qquad (1)$$

The value function is a specific feature of MDPs: the search for an optimal policy can be directly transformed into an optimization problem expressed in terms of value functions. In the literature, there are three efficient methods designed to solve MDPs with a discounted criterion: linear programming (LP), value iteration and policy iteration [2, 3, 29, 31]. We are interested in the LP method which solves MDPs by an optimization approach, in particular a linear programming. In fact, this approach always furnishes the global solutions but with rather long resolution times in practice [31]. This motivates us to study another optimization approach based on the special property of the optimal value function V^*. Indeed, as mentioned above, V^* is a unique solution of the following system of equations:

$$T^*V - V = 0.$$

Therefore, we get the simple idea of directly minimizing the different norms of the Optimal Bellman Residual (OBR), $T^*V - V$. In general, the OBR problems are non-convex and thus solving such a problem by global approaches is very difficult.

We investigate a new and efficient local optimization approach for solving these problems based on DC (Difference of Convex functions) programming and DCA (DC Algorithms). DC programming and DCA were introduced by Pham Dinh Tao in a preliminary form in 1985 and extensively developed since 1994 by Le Thi Hoai An and Pham Dinh Tao (see [15, 16, 25, 26] and the references therein). This work is motivated by the fact that DCA has been successfully applied to many (smooth or nonsmooth) large scale nonconvex programs in various domains of applied sciences, in particular in Machine Learning, for which they provide quite often a global solution and are proved to be more robust and efficient than the standard methods (see, e.g. [8, 9, 12, 14–16, 25–27, 30, 32] and the list of references in [13]).

We consider an optimization formulations of OBR with ℓ_p-norm, in particular ℓ_1-norm. We prove that the ℓ_p-norm problem is a DC program and so DCA can be applied. A so-called DC program is that of minimizing a DC function $f = g - h$ over a convex set with g and h being convex functions. The construction of DCA involves the convex DC components g and h but not the DC function f itself. Moreover, a DC function f has infinitely many DC decompositions $g - h$ which have a crucial impact on the qualities (speed of convergence, robustness, efficiency, globality of computed solutions, and so on) of DCA. The search of a "good" DC decomposition is important from an algorithmic point of views. How to develop an efficient algorithm based on the generic DCA scheme for a practical problem is thus a judicious question to be studied, the answer depends on the specific structure of the problem being considered.

In this paper, our main contribution is threefold. Firstly, we present an ℓ_1-norm optimization formulation of OBR, reformulated as a DC program. Secondly, we propose a suitable DC decomposition for the ℓ_1-norm optimization problem. Finally, we test some numerical experiments in the stationary Garnet problem with small, medium scale settings.

The rest of the paper is organized as follows. In Sect. 2, we describe an optimization formulation of OBR. Section 3 first presents a short introduction of DC programming and DCA and then shows a solution method for the optimization problem by DC programming and DCA. Section 4 reports the numerical results on several test problems which is followed by some conclusions in Sect. 5.

2 Optimization Formulations of Optimal Bellman Residual Problems

As described above, the optimal value function V^* is the fixed point of the operator T^*, we refer here to an optimization approach which aims at directly minimizing the ℓ_p-norm of the OBR over $\mathbb{R}^S$:

$$J_{p,\mu}(V) = ||T^*V - V||_{p,\mu},$$

where $p \in \mathbb{N}, p \geq 1$, μ is the probability distribution over S and $||V||_{p,\mu} = (\sum_{s \in S} \mu(s)|V(s)|^p)^{\frac{1}{p}}$. Obviously, V^* is an optimal solution of $J_{p,\mu}$. In theory, the performance loss resulting from using the policy π greedy with respect to V instead of an optimal policy π^* is bound in terms of the ℓ_p-norm of OBR as follows.

$$||V^* - V^\pi||_{p,\nu} \leq \frac{2}{1 - \gamma} C(\nu, \mu)^{\frac{1}{p}} ||T^*V - V||_{p,\mu}, \tag{2}$$

where $C(\nu, \mu)$ is a constant that measures the concentrability (relative to μ) of the discounted future state distribution (given that the initial state is sampled from ν) of the MDP (see [23, 31] for a precise definition). This result tells us that if the norm of the OBR $J_{p,\mu}$ is well minimized, then the performance of the corresponding greedy policy is close to the optimum.

Generally speaking, the optimization problem of ℓ_p-norm of OBR, $J_{p,\mu}$, is nonconvex, nonsmooth. In general, the optimization problem

$$\min_{V \in \mathbb{R}^S} J_{p,\mu} = \min_{V \in \mathbb{R}^S} ||T^*V - V||_{p,\mu}^p \tag{3}$$

is a DC program (the proof in Sect. 3.2). In this paper, for finding an explicit DC decomposition of (3), we consider the case of a uniform distribution μ over the state space S and $p = 1$. Thus, we describe the optimization formulation to the empirical ℓ_1-norm of OBR in more detail below. When $p = 1$, the empirical ℓ_1-norm optimization formulation of OBR over $\mathbb{R}^S$ would be

$$\min_{V \in \mathbb{R}^S} \left\{ F(V) := \sum_{i=1}^{N_S} |T^*V(s_i) - V(s_i)| \right\}. \tag{4}$$

Remark that the problem (3) and (4) are known to be NP-hard. Moreover, we can use the power of DC programming and DCA for such difficult problems in particular and many (smooth or nonsmooth) nonconvex programs in general.

In Sect. 3, we will prove that (3) is in general a DC program, present a new DC decomposition for (4) and study DCA for solving it.

3 Solution Methods by DC Programming and DCA

Before discussing the algorithms based on DCA for problems (3) and (4), let us introduce briefly DC programming and DCA. A complete study of DC programming and DCA are referred to [15, 16, 25, 26] and the list of references in [13].

3.1 Introduction of DC Programming and DCA

DC Programming and DCA constitute the backbone of smooth/nonsmooth nonconvex programming and global optimization. They address the problem of minimizing a function f which is a difference of convex functions on the whole space $\mathbb{R}^p$. Generally speaking, a DC program takes the form

$$\alpha = \inf \{f(x) := g(x) - h(x) : x \in \mathbb{R}^p\} \quad (P_{dc}) \tag{5}$$

where $g, h \in \Gamma_0(\mathbb{R}^p)$, the set contains all lower semicontinuous proper convex functions on $\mathbb{R}^p$. Such a function f is called DC function, and $g - h$, DC decomposition of f while g and h are DC components of f. The convex constraint $x \in C$ can be incorporated in the objective function of (P_{dc}) by adding the indicator function of C ($\chi_C(x) = 0$ if $x \in C$, $+\infty$ otherwise) to the first DC component g of the DC objective function f.

A DC program (P_{dc}) is called polyhedral DC program when either g or h is a polyhedral convex function (i.e. the sum of a pointwise supremum of a finite collection of affine functions and the indicator function of a nonempty polyhedral convex set). Note that a polyhedral convex function is almost always differentiable, say, it is differentiable everywhere except on sets of measure zero.

Recall that, for $\theta \in \Gamma_0(\mathbb{R}^p)$ and $x \in \text{dom } \theta := \{u \in \mathbb{R}^p | \theta(u) < +\infty\}$, the subdifferent of θ at x, denoted $\partial\theta(x)$, is defined as

$$\partial\theta(x) := \{y \in \mathbb{R}^p : \theta(u) \geq \theta(x) + \langle u - x, y \rangle, \forall u \in \mathbb{R}^p\} \tag{6}$$

which is a closed convex set in $\mathbb{R}^p$. Denote dom $\partial\theta = \{x \in \mathbb{R}^p | \partial\theta(x) \neq \emptyset\}$. It generalizes the concept of derivative in the sense that θ is differentiable at x if and only if $\partial\theta(x)$ is reduced to a singleton set, i.e. $\partial\theta(x) = \{\nabla\theta(x)\}$. Each $y \in \partial\theta(x)$ is called subgradient of θ at x.

The necessary local optimality condition for the primal DC program, (P_{dc}), is

$$\partial h(x^*) \subset \partial g(x^*). \tag{7}$$

This condition (7) is also sufficient for many important classes of DC programs, for example, for polyhedral DC programs, or when function f is locally convex at x^* [16, 25].

A point that x^* satisfies the generalized Kuhn-Tucker condition

$$\partial g(x^*) \cap \partial h(x^*) \neq \emptyset \tag{8}$$

is called a critical point of $g - h$. It follows that if h is polyhedral convex, then a critical point of $g - h$ is almost always a local solution to (P_{dc}).

Based on local optimality conditions and duality in DC programming, the idea of DCA is quite simple: each iteration k of DCA approximates the concave part $-h$ by its affine majorization corresponding to taking $y \in \partial h(x^k)$ and minimizes the resulting convex function (that is equivalent to determining $x^{k+1} \in \partial g^*(y^k)$).

Generic DCA scheme
Initialization: Let $x^0 \in \mathbb{R}^p$ be a best guess, $k := 0$.
Repeat

- Calculate $y^k \in \partial h(x^k)$
- Calculate $x^{k+1} \in \text{argmin}\{g(x) - h(x^k) - \langle x - x^k, y^k \rangle : x \in \mathbb{R}^p\}$ (P_k)
- $k = k + 1$

Until convergence of $\{x^k\}$.

Convergence properties of the DCA and its theoretical basis are described in [16, 25, 26]. However, it is worthwhile to report the following properties that are useful in the next section (for simplify, we omit here the dual part of these properties):

- DCA is a descent method *without linesearch* (the sequence $\{g(x^k) - h(x^k)\}$ is decreasing) but with global convergence (i.e. it converges from arbitrary starting point).
- If $g(x^{k+1}) - h(x^{k+1}) = g(x^k) - h(x^k)$, then x^k is a critical point of $g - h$. In this case, DCA terminates at kth iteration.
- If the optimal value α of problem (P_{dc}) is finite and the infinite sequence $\{x^k\}$ is bounded, then every limit point x^* of this sequence is a critical point of $g - h$.
- DCA has a linear convergence for general DC programs. Especially, for polyhedral DC programs the sequence $\{x^k\}$ contains finitely many elements and the algorithm convergences to a solution in a finite number of iterations.

In the past years, DCA has been successfully applied in several works of various fields among them learning machines, financial optimization, supply chain management, etc. (see [14, 17–22, 24, 27] and the list of references in [13]).

3.2   The ℓ_p-norm Formulation of OBR Is a DC Program

In this subsection, we prove that for any $p \in \mathbb{N}, p \geq 1$, the ℓ_p-norm formulation (3) is certainly a DC program. Through the paper, for each $i \in \{1, \dots, N_S\}$, let f_i be a function from $\mathbb{R}^S$ to $\mathbb{R}$ defined by

$$f_i(V) = T^* V(s_i) - V(s_i)$$

$$= \max_{j=1,\dots,N_A} \left(\mathcal{R}(s_i, a_j) + \gamma \sum_{s' \in S} P(s'|s_i, a_j) V(s') \right) - V(s_i).$$

Consequently, the problem (3) can be expressed as

$$\min \left\{ J_{p,\mu}(V) = \sum_{i=1}^{N_S} \mu(s_i)|f_i(V)|^p : V \in \mathbb{R}^S \right\}. \qquad (9)$$

Theorem 1 *For any $p \in \mathbb{N}, p \geq 1, J_{p,\mu}$ is a DC function.*

Proof We have

$$J_{p,\mu}(V) = \sum_{i=1}^{N_S} \mu(s_i)|f_i(V)|^p. \qquad (10)$$

Since for each $s_i \in S$ and $a_j \in \mathcal{A}$, the function

$$\mathcal{R}(s_i, a_j) + \gamma \sum_{s' \in S} P(s'|s_i, a_j) V(s') - V(s_i)$$

is linear in V, the function f_i, a finite maximum of linear functions, is polyhedral convex.

Due to the fact that

$$|f_i| = f_i^+ - f_i^-,$$

where $f_i^+ := \max\{0, f_i\}, f_i^- := \max\{0, -f_i\}$ is also nonnegative polyhedral functions, the function $|f_i|$ is a DC function.

As both a finite product and a weighted sum of DC functions whose DC components are nonnegative are also a DC function, then this proof is complete. □

Since $J_{p,\mu}$ is a DC function, the problem (3) is a DC program. In general, DCA can be applied to solve (3). In fact, in order to get an explicit DC decomposition, we will consider an interesting case ($p = 1$), the ℓ_1-norm problem as mentioned in Sect. 2. Recall that the empirical ℓ_1-norm optimization formulation of OBR over $\mathbb{R}^S$ is

$$\min_{V \in \mathbb{R}^S} F(V) = \min_{V \in \mathbb{R}^S} \sum_{i=1}^{N_S} |f_i(V)|.$$

Unlike the proof of Theorem 1, we consider the particular problem (4) with ℓ_1-norm instead of ℓ_p-norm. We see that for $i \in \{1, \dots, N_S\}$,

$$|f_i(V)| = 2f_i^+(V) - f_i(V)$$

and $f_i^+(V), f_i(V)$ are polyhedral convex. It follows that the following DC formulation of (4) is of the form:

$$\min \left\{ F(V) := G(V) - H(V) : V \in \mathbb{R}^S \right\}, \tag{11}$$

where

$$G(V) = \sum_{i=1}^{N_S} 2f_i^+(V), \ H(V) = \sum_{i=1}^{N_S} f_i(V).$$

It is clear that the function F is a polyhedral DC function with both polyhedral convex DC components G and H. Hence, the problem (11) is a polyhedral DC program. DCA enjoys interesting properties in this case.

In the sequel, we will present how to apply DCA to the empirical ℓ_1-norm formulation of OBR, specifically the ℓ_1-norm problem (11).

3.3 DCA for Solving the ℓ_1-norm Problem (11)

According to the generic DCA scheme, at each iteration k, after computing a subgradient $w^k \in \partial H(V^k)$, the calculation of V^{k+1} is reduced to solve the convex program:

$$\min \left\{ \sum_{i=1}^{N} 2f_i^+(V) - \langle w^k, V \rangle : V \in \mathbb{R}^{N_S} \right\}. \tag{12}$$

Solving the subproblem (12) amounts to solving the following linear program:

$$\begin{cases} \min \sum_{i=1}^{N_S} 2t_i - \langle w^k, V \rangle, \\ \text{s.t. } V \in \mathbb{R}^{N_S}, \\ t_i \geq 0, \forall i = 1, \dots, N_S, \\ t_i \geq R(s_i, a_j) + \gamma \langle P(\cdot | s_i, a_j) - e_i, V \rangle, \\ \quad \forall i = 1, \dots, N_S, \forall j = 1, \dots, N_A. \end{cases} \tag{13}$$

Compute ∂H: By the definition of H, we have

$$\partial H(V) = \sum_{i=1}^{N_S} \partial f_i(V)$$

$$= \sum_{i=1}^{N_S} \partial \left[\max_{j=1,\dots,N_A} \left(R(s_i, a_j) + \gamma \sum_{s' \in S} P(s' | s_i, a_j) V(s') \right) - V(s_i) \right]$$

$$= \sum_{i=1}^{N_S} \text{co} \left\{ \gamma \mathcal{P}(\cdot | s_i, a_{ji}) - e_i \ : \ a_{j_i} \in I_i(V) \right\}, \tag{14}$$

where co$\{\mathcal{X}\}$ denotes the convex hull of a set of points $\mathcal{X}$; e_i is the ith unit vector in $\mathbb{R}^{N_S}$; the vector $\mathcal{P}(\cdot | s_i, a_j) = \left(\mathcal{P}(s_1 | s_i, a_j), \dots, \mathcal{P}(s_{N_S} | s_i, a_j) \right)^\top \in \mathbb{R}^{N_S}$ and

$$I_i(V) = \text{argmax}_{a' \in \mathcal{A}} \left\{ \mathcal{R}(s_i, a') + \gamma \langle \mathcal{P}(\cdot | s_i, a'), V \rangle \right\}.$$

Let us define $t = (t_1, \dots, t_{N_S})^\top \in \mathbb{R}^{N_S}$. From the generic DCA scheme described in Sect. 3.1, DCA applied to (11) is given by Algorithm ℓ_1-DCA.

Algorithm ℓ_1-DCA (DCA for solving (11))

Initialization: Let ε be a sufficiently small positive number. Let $V^0 \in \mathbb{R}^d$. Set $k = 0$.

repeat
    1. Compute $w^k \in \partial H(V^k)$ by using (14).
    2. Compute (V^{k+1}, t^{k+1}), an optimal solution of the linear program (13).
    3. $k = k + 1$.
until $|F(V^{k+1}) - F(V^k)| \leq \varepsilon(|F(V^k)| + 1)$ or $||V^{k+1} - V^k||_2 \leq \varepsilon \max(1, ||V^k||_2)$ or $F(V^{k+1}) = 0$.

According to the properties of convergence of DCA mentioned above and observing that H is a polyhedral convex function, we deduce the following convergence properties of ℓ_1-DCA.

Theorem 2 *Convergence properties of ℓ_1-DCA*
(i) ℓ_1-DCA generates the sequence $\{V^k\}$ containing finitely many elements such that the sequence $\{F(V^k)\}$ is decreasing.
(ii) After a finite number of iterations, the sequence $\{V^k\}$ converges to V^* that is a critical point of (11). Moreover, if $I_i(V^*)$ is a singleton set for all $i \in \{1, \dots, N_S\}$, then V^* is a local minimizer to (11).
(iii) If $F(V^*) = 0$, then V^* is a global minimizer to problem (11).

Proof (i) This property is direct consequences of the convergence properties of polyhedral DC programs.
(ii) Since H is a polyhedral convex function, the necessary local optimality condition $\partial H(V^*) \subset \partial G(V^*)$ is also sufficient. This inclusion holds when $I_i(V^*)$ is a singleton set for all $i \in \{1, \dots, N_S\}$.
(iii) It is evident due to the contracting operator T^*.
The proof is then complete. $\qquad \square$

4 Numerical Experiments

In our experiments, ℓ_1-DCA was tested on stationary Garnet problems taken from
[1]. For a fair comparison, we compared ℓ_1-DCA with LP method as mentioned in
Sect. 1 with the optimization approach via solving a mathematical programming and
thus, other approaches were considered as another standard. All experiments were
implemented in the MATLAB R2013b and performed on a PC Intel(R) Core(TM)
i5-3470 CPU, 3.20GHz of 8GB RAM. Now, we give a brief presentation of Garnet
problems. A stationary Garnet problem with finite MDP is characterized by 3 parameters: Garnet(N_S, N_A, N_B). The parameters N_S and N_A are the number of states and
actions respectively, and N_B is a branching factor specifying the number of next states
for each state-action pair. In these experiments, a specific type of Garnets presenting
a topological structure relative to real dynamical systems [28] is considered. Those
systems are generally multidimensional state spaces MDPs where an action leads to
different next states close to each other. The fact that an action leads to close next
states can model the noise in a real system for instance. Hence, problems such as the
highway simulator [11], the mountain car or the inverted pendulum (possibly dis-
cretized) are particular cases of this type of Garnets. The state space S is considered
as $S = \{s_i\}_{i=1,\dots,N_S}$ where each state $s_i = (s_i^{(j)})_{j=1,\dots,m}$ is an m−tuple ($m = 2$) and
each component $s_i^{(j)}$ is chosen out of all integer numbers between 1 and x_i. Thus, the
number of states N_S is $\prod_{i=1}^{m} x_i$. The number of actions N_A is set to 4. For each state
action couple (s, a), we can move to any next state $s' \in S$ and thus, $N_B = N_S$. The
probability of going to each next state s' is generated by partitioning the unit interval
at $N_B - 1$ cut points selected randomly. For each couple (s, a), the reward $\mathcal{R}(s, a)$ is
drawn uniformly between -1 and 1. The discount factor γ is set to 0.95 in all experiments. For each size of N_S, we drew independently 10 Garnet problems $\{G_p\}_{1 \le p \le 10}$
described above.

In addition, for each experiment, our algorithm was started from the point $V^0 = 0 \in \mathbb{R}^{N_S}$ and taken the same stop criterion with the tolerance $\varepsilon = 10^{-4}$. CPLEX 12.6
was used for solving linear programs. We are interested in the following aspects to
examine the effectiveness of the proposed algorithms: the CPU time (in seconds),
the value of objective functions $||F(V)||_1$ and the policy π greedy with respect to
the output V of each algorithm. Specifically, for each Garnet problem, we use Diff to
denote the difference between the given policy and the optimal policy i.e. the number
of states on which the given policy generated by an efficient iterative method, Policy
Iteration, differs from the optimal policy obtained by LP method and ℓ_1-DCA. For
each size of N_S, the experiment was conducted over 10 Garnet problems. The results
were reported in Table 1 by averaging over these 10 Garnet problems with different
small/medium sizes of N_S, in particular $N_S \in \{25, 100, 225, 400, 625, 900, 1225\}$.
From the suitable starting point V^0, ℓ_1-DCA furnished the optimal solution after
exactly one iteration. This shows that the better the starting point is, the more efficient
our proposed algorithm would be.

Table 1 Comparative results of ℓ_1-DCA and LP on all 10 Garnet experiments for each N_S

N_S	CPU (s)		Val. Obj. $\|F(V)\|_1$		Diff	
	LP	ℓ_1-DCA	LP	ℓ_1-DCA	LP	ℓ_1-DCA
25	0.0125	0.0001	2.52e-14	5.04e-14	0	0
100	0.1310	0.0936	1.39e-13	3.87e-13	0	0
225	0.6490	0.4758	4.00e-13	1.33e-12	0	0
400	2.0842	1.9391	8.42e-13	3.14e-12	0	0
625	7.0793	6.9218	1.48e-12	6.25e-12	0	0
900	99.073	93.621	2.36e-12	1.06e-11	0	0
1225	242.42	229.72	3.38e-12	1.67e-11	0	0

Comments on numerical results

We observe from the numerical results that concerning the value of objective function, both algorithms have the comparable results in all cases. In terms of the quality of solutions, ℓ_1-DCA is very efficient: it furnished an optimal solution V with the accuracy $\|F(V)\|_1$ less than 10^{-10} in all experiments. Moreover, observing the difference Diff, we see that the policy greedy with respect to the optimal solution V of ℓ_1-DCA is exactly the same as that of the LP method and the exact policy as well. As a result, the proposed algorithm provides the good policy for the MDP problems. Generally speaking, our algorithm works well on these Garnet problems. Regarding the CPU time, ℓ_1-DCA is more effective: in all cases, it is faster than the LP method.

5 Conclusion

We have intensively studied DC programming and DCA for solving MDP problems. Based on the optimization approach, an OBR minimization problem is considered for which the DCA based approach can be used. By exploiting the special structure of the considered problems, we have developed a DCA scheme with suitable DC decompositions. Furthermore, from numerical experiments on the Garnet problem, the comparative results show the efficiency of our proposed approach, in particular ℓ_1-DCA in the small/medium problems. In the future, by our proposed approach, we are interested in the large scale problem without knowing the perfect dynamic (a model of the state transition probability function and reward function) of the underlying MDP as well as various problems modeled in terms of an MDP.

References

1. Archibald, T., McKinnon, K., Thomas, L.: On the generation of markov decision processes. J. Oper. Res. Soc. **46**(3), 354–361 (1995)
2. Bellman, R.E.: Dynamic Programming. Princeton University Press, Princeton (1957)
3. Bertsekas, D.P. (ed.): Dynamic Programming: Deterministic and Stochastic Models. Prentice-Hall Inc, Upper Saddle River (1987)
4. Bertsekas, D.P. (ed.): Introduction to Reinforcement Learning. MIT Press, Cambridge (1998)
5. Bertsekas, D.P., Tsitsiklis, J.N. (eds.): Neuro-Dynamic Programming. Athena Scientific (1996)
6. Boutilier, C.: Knowledge representation for stochastic decision processes. In: Wooldridge, M.J., Veloso, M. (eds.) Artificial Intelligence Today, Lecture Notes in Computer Science, vol. 1600, pp. 111–152. Springer, Berlin (1999). http://dx.doi.org/10.1007/3-540-48317-9_5
7. Boutilier, C., Dean, T., Hanks, S.: Decision-theoretic planning: structural assumptions and computational leverage. J. Artif. Intell. Res. **11**, 1–94 (1999)
8. Chan, A.B., Vasconcelos, N., Lanckriet, G.R.G.: Direct convex relaxations of sparse svm. In: Langley, P. (ed.) Proceedings of the 24th International Conference on Machine Learning, pp. 145–153. ACM, New York, NY, USA (2007)
9. Collobert, R., Sinz, F.H., Weston, J., Bottou, L.: Trading convexity for scalability. In: ICML. pp. 201–208 (2006)
10. Kaelbling, L.P., Littman, M.L., Moore, A.W.: Reinforcement learning: a survey. J. Artif. Intell. Res. **4**, 237–285 (1996)
11. Klein, E., Geist, M., Piot, B., Pietquin, O.: Inverse reinforcement learning through structured classification. In: Pereira, F., Burges, C., Bottou, L., Weinberger, K. (eds.) Advances in Neural Information Processing Systems 25, pp. 1007–1015. Curran Associates, Inc. (2012)
12. Krause, N., Singer, Y.: Leveraging the margin more carefully. In: ICML'04: Proceedings of the twenty-first international conference on Machine learning. pp. 63. ACM Press, New York, NY, USA (2004)
13. Le Thi, H.A.: DC programming and DCA (2012). http://www.lita.univ-lorraine.fr/~lethi
14. Le Thi, H.A., Moeini, M.: Long-short portfolio optimization under cardinality constraints by difference of convex functions algorithm. J. Optim. Theory Appl. **161**(1), 199–224 (2014)
15. Le Thi, H.A., Pham Dinh, T.: Solving a class of linearly constrained indefinite quadratic problems by D.C. algorithms. J. Glob. Optim. **11**(3), 253–285 (1997)
16. Le Thi, H.A., Pham Dinh, T.: The DC (difference of convex functions) programming and DCA revisited with DC models of real world nonconvex optimization problems. Ann. Oper. Res. **133**(1–4), 23–46 (2005)
17. Le Thi, H.A., Vo, X.T., Pham Dinh, T.: Robust Feature Selection for SVMs under Uncertain Data. In: Perner, P. (ed.) Advances in Data Mining. Applications and Theoretical Aspects, pp. 151–165. Springer, Berlin (2013)
18. Le Thi, H., Pham Dinh, T., Le, H., Vo, X.: Dc approximation approaches for sparse optimization. Eur. J. Oper. Res. **244**(1), 26–46 (2015)
19. Le Thi, H.A., Le, H.M., Pham Dinh, T.: Feature selection in machine learning: an exact penalty approach using a difference of convex function algorithm. Mach. Learn. **101**(1–3), 163–186 (2015)
20. Le Thi, H.A., Nguyen, M.C., Pham Dinh, T.: A dc programming approach for finding communities in networks. Neural Comput. **26**(12), 2827–2854 (2014)
21. Le Thi, H.A., Vo, X.T., Pham Dinh, T.: Feature selection for linear SVMs under uncertain data: Robust optimization based on difference of convex functions algorithms. Neural Netw. **59**, 36–50 (2014)
22. Le Thi, H., Nguyen, M.: Self-organizing maps by difference of convex functions optimization. Data Min. Knowl. Discov. **28**(5–6), 1336–1365 (2014)
23. Munos, R.: Performance bounds in L_p norm for approximate value iteration. SIAM J. Control Optim. (2007)
24. Pham Dinh, T., Le, H.M., Le Thi, H.A., Lauer, F.: A difference of convex functions algorithm for switched linear regression. IEEE Trans. Autom. Control **59**(8), 2277–2282 (2014)

25. Pham Dinh, T., Le Thi, H.A.: Convex analysis approach to d.c. programming: theory, algorithms and applications. Acta Math. Vietnam. **22**(1), 289–355 (1997)

26. Pham Dinh, T., Le Thi, H.A.: DC optimization algorithms for solving the trust region subproblem. SIAM J. Optim. **8**(2), 476–505 (1998)

27. Pham Dinh, T., Le Thi, H.A.: Recent advances in DC programming and DCA. In: Nguyen, N.T., Le Thi, H.A. (eds.) Transactions on Computational Intelligence XIII, vol. 8342, pp. 1–37. Springer, Berlin (2014)

28. Piot, B., Geist, M., Pietquin, O.: Differene of convex functions programming for reinforcement learning. In: Ghahramani, Z., Welling, M., Cortes, C., Lawrence, N.D., Weinberger, K.Q. (eds.) Advances in Neural Information Processing Systems 27, pp. 2519–2527. Curran Associates, Inc. (2014)

29. Puterman, M.L. (ed.): Markov Decision Processes: Discrete Stochastic Dynamic Programming. Wiley, New York (1994)

30. Schüle, T., Schnörr, C., Weber, S., Hornegger, J.: Discrete tomography by convex-concave regularization and d.c. programming. Discret. Appl. Math. **151**, 229–243 (2005)

31. Sigaud, O., Buffet, O. (eds.): Markov Decision Processes in Artificial Intelligence. Wiley, IEEE Press (2010)

32. Yin, P., Lou, Y., He, Q., Xin, J.: Minimization of $L_1 - L_2$ for compressed sensing. SIAM J. Sci. Comput. (to appear)

Part II
Models for ICT Applications

A Method for Transforming TimeER Model-Based Specification into Temporal XML

Quang Hoang, Tinh Van Nguyen, Hoang Lien Minh Vo
and Truong Thi Nhu Thuy

Abstract Time is an attribute of the phenomenon in the real world because the information we receive always changes with time. In this paper, we propose a method to transform specifications in TimeER model into XML schema by transforming the temporal entities, temporal attributes and temporal relationships in a TimeER model to XML schema. Our transformation approach is based on the improvement of the transformation methods from ER model to XML schema and the addition of the transforming rules for the temporal components of the TimeER model.

Keywords Temporal database model · TimeEr model · Temporal XML document · Temporal XML schema

1 Introduction

The value of database is not the size of data; that is the amount of information which can be retrieved from it. Therefore, the information of the past or present is very important. Temporal database was researched to store information of data at different times. But when designing the temporal conceptual database, we must

Q. Hoang (✉) · T. Van Nguyen · H.L.M. Vo
Hue University of Sciences, 77 Nguyen Hue, Hue City, Vietnam
e-mail: hquang@hueuni.edu.vn

T. Van Nguyen
e-mail: nvtinh@gmail.com

H.L.M. Vo
e-mail: minhvhl@gmail.com

T.T.N. Thuy
Hue University College of Foreign Languages, 57 Nguyen Khoa Chiem,
Hue City, Vietnam
e-mail: nhuthuytt@yahoo.com

© Springer International Publishing Switzerland 2016
T.B. Nguyen et al. (eds.), *Advanced Computational Methods
for Knowledge Engineering*, Advances in Intelligent Systems
and Computing 453, DOI 10.1007/978-3-319-38884-7_5

choose the simple form that can store time state. Using the ER models for designing database will encounter difficulties because the requirement is to support the temporal component, while the traditional ER model cannot meet those requirements. To solve the problem of temporal conceptual schema design, the research community has developed many different temporal ER models such as TERM, RAKE, MOTAR, TEER, STEER, ERT, TER, TempEER, TempRT, TERC+, and TimeER [1]. In particular, the TimeER model which was proposed by Gregersen and Jensen [1] is one of the relatively updated temporal conceptual model allowing supporting quite fully the temporal component and is commonly used today. This model was developed based on the EER model [1]—a version of extended ER proposed by Elmasri and Navathe.

Currently, XML (eXtensible Markup Language) is quite popular; it is based on HTML's simplicity and ease to use of and SGML's complexity and multi-function. XML is designed to allow computers to exchange documents together via the Web without losing semantics of data. The XML documents are suitable for presenting the structured and semi-structured data. The structure and syntax of XML is quite simple, but their application has great significance. Besides, to present the temporal component of real world, the XML documents has been extended towards supporting the temporal components, called the Temporal XML [2, 3]. Accordingly, we need to build a method to design Temporal XML schema from the temporal conceptual database models, namely, TimeER model, so that this transforming can preserve information and avoid data redundancy. TimeER model is an extension of EER model. Although there have been proposals for the transforming methods from the traditional ER model into XML schema [4–8], the extending in temporal database has not been mentioned in previous study. Therefore, this paper proposes a method to transform specifications from TimeER model into Temporal XML.

Accordingly, this paper is organized as follows: in the next section, we present an overview of components of TimeER model and Temporal XML schema. Section 3 is about the rules of transforming from TimeER model into Temporal XML. The last one is the conclusion.

2   An Overview of the TimeER Model and Temporal Schema

2.1   TimeER Model

TimeER model is an extension of EER model by allowing quite fully support the temporal components in comparing to other temporal ER models. Some temporal aspects that this model supported are the lifespan of an entity, the valid time of a fact, and the transaction time of an entity or a fact [1, 9–11].

This model has a convention that temporal aspects of the entities in the database can be either the lifespan (LS), or the transaction time (TT), or both the lifespan and the transaction time (LT). The temporal aspects of the attributes of an entity can be either the valid time (VT), or the transaction time (TT), or both the valid time and the transaction time (BT). Besides, because a relationship type can be seen as an entity type or an attribute, consequentially, the designer can define the temporal aspects supported with this relationship type, if necessary.

Components that support the temporal aspect:

- Entity types: In TimeER schema, each entity type can support either the lifespan, or the transaction time, or both the lifespan and the transaction time, by placing a LS, or a TT, or a LT, respectively, behind the entity type name.
- Attributes: If an attribute supports the valid time, or the transaction time, or both of them are captured, this is indicated by placing a VT, or a TT, or a BT (BiTemporal), respectively, behind the attribute name.
- Relationships: For each relationship type, it can be decided by the database designer whether or not to capture the temporal aspects of the relationships of the relationship type. Specifically, the relationship can be seen as an entity type (using the symbols LS, TT or LT) or an attribute (using the symbols VT, TT or BT). If some temporal aspect is captured for a relationship type we term it temporal; otherwise, it is called non-temporal (Fig. 1).

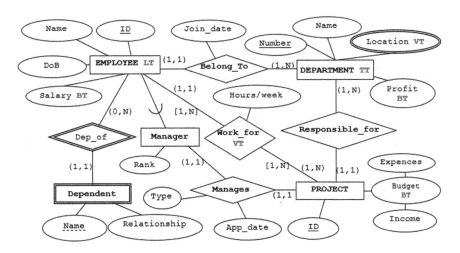

Fig. 1 An example of TimeER model [1]

2.2   Temporal XML Schema

Temporal XML schema is an XML schema which allows adding the temporal attribute declared in temporal components. Similarly, add two timestamp attributes VTs and VTe for the supporting valid time; add timestamp attributes TTs and TTe for transaction time and add VTs, VTe, TTs, TTe for both temporal supports [2, 3, 12–15].

Let's consider an instance of Temporal XML with attribute *Salary* which supports valid time (for storage salary history of employee).

```xml
<?xml version="1.0" encoding="utf-8"?>
  <xs:schema xmlns:xs="http://www.w3.org/2001/XMLSchema">
    <xs:element name="EMPLOYEE">
    <xs:complexType>
      <xs:sequence>
        <xs:element name="ID" type="xs:string"/>
        <xs:element name="Name" type="xs:string"/>
        <xs:element name="DoB" type="xs:date"/>
        <xs:element name="Salary" minOccurs="1" maxOccurs="unbounded">
          <xs:complexType>
            <xs:complexContent>
              <xs:extension base="xs:long">
                <xs:attribute name="VTs" type="xs:date"/>
                <xs:attribute name="VTe" type="xs:date"/>
              </xs:extension>
            </xs:complexContent>
          </xs:complexType>
        </xs:element>
      </xs:sequence>
    </xs:complexType>
    <xs:key name="keyEmp">
        <xs:selector xpath=".//EMPLOYEE"/>
        <xs:field xpath="ID"/>
    </xs:key>
    </xs:element>
  </xs:schema>
```

Element *Salary* is declared complex type. Declare `<xs:complexContent>` indicates that element *Salary* can refer to data values only. `<xs:extension base="xs:long">` indicates the data values such as "long". Two lines `<xs:attribute name="VTs" type="xs:date"/>` and `<xs:attribute name="VTe" type="xs:date"/>` declare two timestamp attributes.

If the declaration of the datatype of the element/attribute is not concerned, the temporal XML schema can be expressed as a tree as follows:

```
EMPLOYEE(ID, DoB, Name, Salary+)
  Salary(VTs,VTe)
  KEY(EMPLOYEE.ID)
```

Or it can be represented as Fig. 2.

Fig. 2 The hierarchical of
Employee structure

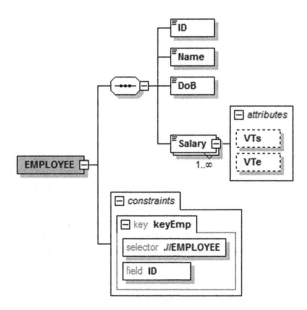

3   Transformation of TimeER Model into Temporal XML Schema

Franceschet et al. [6] have proposed a method to allow mapping almost all the components in the ER model into XML Schema. The study introduces the transformation rules for the entity types, attributes and relationships. When transforming the traditional ER model into XML schema, Massimo Franceschet proposed two different ways: (1) a maximum connectivity nesting that minimizes the number of schema constraints used in the mapping of the conceptual schema reducing the validation overhead. (2) a maximum depth nesting that lowers the number of (expensive) joining operations necessary for the reconstruction of the information at query time using the mapped schema. They proposed a graph-theoretic linear-time algorithm to find a maximum connectivity nesting and proved that finding a maximum depth nesting is a completed NP problem.

With the first proposal, the authors have introduced methods for basic mapping. However, they have not come up with methods for converting the nested multi-valued composite attributes [5]. The studies proposed a method to map the nested multi-valued composite attributes of an entity type in ER model.

The method for transforming TimeER model into Temporal XML is presented in this paper is extension of the method for transforming ER model into Temporal XML. This method adds the rules to transform the components of temporal ER model, including: the temporal entity type, the temporal attribute of entity type, the temporal relationship, and the temporal attribute of relationship.

3.1   The Transformation Rules

Rule 1: Transformation of temporal entity type

If: The entity type E^* is the entity type which the temporal support of E is indicated by an asterix (*) and key is KE.

Then:

- Create a complex element E.
- Create a complex element S_E nesting in the complex element E, add two constraint attributes minOccurs = "0" and maxOccurs = "unbounded"
- Create a complex element W_E nesting in S_E and add the temporal attribute correspond with timestamp * in Table 1, simultaneously create the attributes corresponding with the key attributers of entity type E.
- Key of the complex element W_E includes the key KE of entity type E and the partial key T' of timestamp * which are shown in Table 1 (Table 2).

Example 1 TimeER model has an entity type EMPLOYEE which supports lifespan shown in Fig. 3. In application of Rule 1, the result of transformation is Temporal XML schema shown in Fig. 4.

The illustration of the database corresponding to the TimeER model is shown in Fig. 5.

Table 1 Abbreviation used for temporal support of entity types and relationship types

* = LS	$T = \{LSs, LSe\}$ and $T' = \{LSs\}$
* = TT	$T = \{TTs, TTe\}$ and $T' = \{TTs\}$
* = LT	$T = \{LSs, LSe, TTs, TTe\}$ and $T' = \{LSs, LSe, TTs\}$

Table 2 Transformation of the temporal entity type

LS	`E(KE,S_E*)`
	`S_E(W_E)`
	`W_E(LSs, LSe, KE)`
	`KEY(E.KE), KEY(W_E.LSs,W_E.KE)`
	`CHECK(W_E.KE = E.KE)`

TT	`E(KE,S_E*)`
	`S_E(W_E)`
	`W_E(TTs, TTe, KE)`
	`KEY(E.KE), KEY(W_E.TTs,W_E.KE)`
	`CHECK(W_E.KE = E.KE)`

LT	`E(KE,S_E*)`
	`S_E(W_E)`
	`W_E(LSs, LSe, TTs, TTe, KE)`
	`KEY(E.KE), KEY(W_E.LSs, W_E.LSe, W_E.TTs, W_E.KE)`
	`CHECK(W_E.KE = E.KE)`

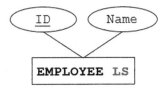

Fig. 3 The TimeER model

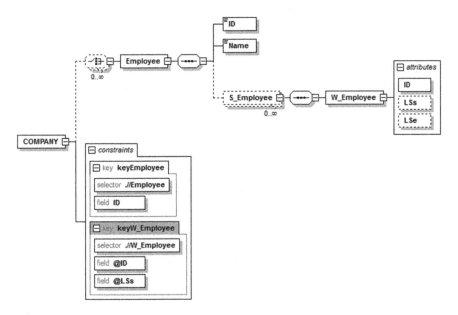

Fig. 4 The result of transformation into Temporal XML schema

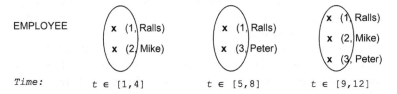

Fig. 5 Database of TimeER model at three times

Corresponding XML document of Temporal XML schema as Fig. 6.

Rule 2: Transformation of temporal attributes of an entity type

If: The entity E has an attribute $A*$ (the attribute A is supported with the temporal aspect *)

Fig. 6 XML document of
Temporal XML schema in
Fig. 4

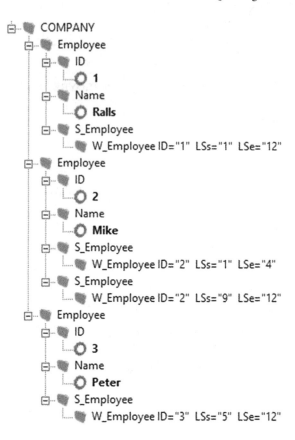

Then:

- Create the complex element *E*.
- Create a complex element *S_A* which is nested in the complex element *E*, simultaneously add two constraint attributes minOccurs = "0" and maxOccurs = "unbounded".
- Create a complex element *W_A* which is nested in *S_A* and add the temporal attributes which correspond to the temporal aspect * shown in Table 3, simultaneously add the attributes which correspond to the key attributes of the entity type *E*.

Table 3 Set of timestamp attributes support for of entity types and relationship types

* = VT	$T = \{VTs, VTe\}$ and $T' = \{VTs\}$
* = TT	$T = \{TTs, TTe\}$ and $T' = \{TTs\}$
* = BT	$T = \{VTs, VTe, TTs, TTe\}$ and $T' = \{VTs, VTe, TTs\}$

Table 4 Transformation of temporal attribute

VT	E(KE,S_A*) S_A(W_A) W_A(VTs, VTe, KE) KEY(E.KE), KEY(W_A.VTs, W_A.KE) CHECK(W_A.KE = E.KE)
TT	E(KE,S_A*) S_A(W_A) W_A(TTs, TTe, KE) KEY(E.KE), KEY(W_A.TTs, W_A.KE) CHECK(W_A.KE = E.KE)
BT	E(KE,S_A*) S_A(W_A) W_A(VTs, VTe, TTs, TTe, KE) KEY(E.KE), KEY(W_A.VTs, W_A.VTe, W_A.TTs, W_A.KE) CHECK(W_A.KE = E.KE)

Fig. 7 Temporal attribute of
an entity type

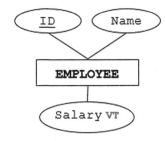

- The primary key of the complex element W_A is created by combining the key KE of the entity E with the partial key T' of the temporal aspect * shown in Table 3 (Table 4).

Example 2 TimeER model has an entity type EMPLOYEE and attribute Salary which supports the valid time shown in Fig. 7. When Rule 2 is applied, the result of transformation is shown in Fig. 8.

The illustration of the database corresponding to the TimeER model is shown in Fig. 9.

Then the data which represent in XML document corresponding with XML schema is shown in Fig. 10. For example, at the time of VTs is 1 and VTe is 8, the salary of Ralls is 500; at the time of VTs is 9 and VTe is 12 then the salary of Ralls is 800, the result of transformation is represented in XML document.

Rule 3. Transformation of temporal relationship

If: The relationship $R*$ between two entity types $E1$ and $E2$ (the relationship R is supported with the temporal aspect *)

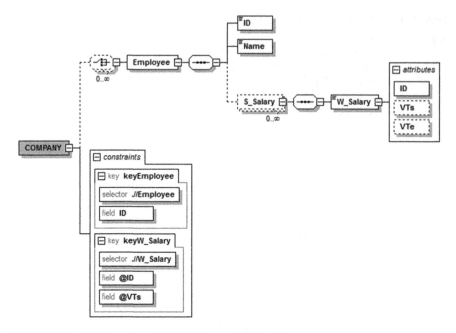

Fig. 8 The result of transformation into temporal XML schema

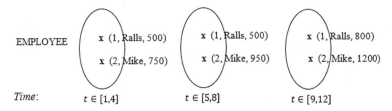

Fig. 9 Database of TimeER model at three times

Then:

- Create a complex element S_R which is nested in R, and add two constraint attributes minOccurs = "0" and maxOccurs = "unbounded".
- Create the complex element W_R which is nested in S_R and add the timestamps corresponding with the temporal aspect * shown in Table 1 or 3, simultaneously add the attributes corresponding with the key attributes of the entity type in relationship R.
- The primary key constraint for the complex element W_R is created by combining the key of the entity types in relationship R and the partial key T' of the temporal aspect * shown in Table 1 or 3.

Fig. 10 XML document of
Temporal XML schema in
Fig. 8

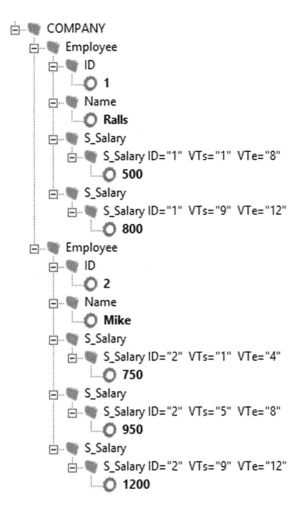

Because this relationship can be only many-to-many binary relationship (because when transformed relationship $R*$, the maximum cardinality constraints of the entity types in the relationship must be changed into N), so that we present the transforming rule of the many-to-many temporal relationship shown in Table 5.

Let us consider LS, the two transformations are equivalent in terms of the number of the used constraints. Nesting of the element for one entity into the element for another one is never possible in the mappings of many-to-many relationships; thus, nest and flat mappings coincide in cases.

Table 5 Transformation of the many-to-many temporal relationship $E_1 \leftarrow (0,N) \rightarrow R \leftarrow (0,N) \rightarrow E_2$

*		R into E_1	R into E_2

LS	Flat	E_1(KE$_1$, R*)
		  R(KE$_2$, S_R*)
		    S_R(W_R)
		      W_R(LSs, LSe, KE$_1$, KE$_2$)
		E_2(KE$_2$)
		KEY(E_1.KE$_1$), KEY(E_2.KE$_2$),
		KEY(R.KE$_2$), KEY(W_R.LSs,
		W_R.KE$_1$, W_R.KE$_2$)
		KEYREF(R.KE$_2$ → E_2.KE$_2$)
		CHECK(W_R.KE$_1$ = E_1.KE$_1$)
		CHECK(W_R.KE$_2$ = E_2.KE$_2$)
			E_2(KE$_2$, R*)
			  R(KE$_1$, S_R*)
			    S_R(W_R)
			      W_R(LSs, LSe, KE$_2$, KE$_1$)
			E_1(KE$_1$)
			KEY(E_1.KE$_1$), KEY(E_2.KE$_2$),
			KEY(R.KE$_1$), KEY(W_R.LSs, W_R.KE$_2$,
			W_R.KE$_1$)
			KEYREF(R.KE$_1$ → E_1.KE$_1$)
			CHECK(W_R.KE$_2$ = E_2.KE$_2$)
			CHECK(W_R.KE$_1$ = E_1.KE$_1$)

TT		E_1(KE$_1$, R*)
		  R(KE$_2$, S_R*)
		    S_R(W_R)
		      W_R(TTs, TTe, KE$_1$, KE$_2$)
		E_2(KE$_2$)
		KEY(E_1.KE$_1$), KEY(E_2.KE$_2$),
		KEY(R.KE$_2$), KEY(W_R.TTs,
		W_R.KE$_1$, W_R.KE$_2$)
		KEYREF(R.KE$_2$ → E_2.KE$_2$)
		CHECK(W_R.KE$_1$ = E_1.KE$_1$)
		CHECK(W_R.KE$_2$ = E_2.KE$_2$)
			E_2(KE$_2$, R*)
			  R(KE$_1$, S_R*)
			    S_R(W_R)
			      W_R(TTs, TTe, KE$_2$, KE$_1$)
			E_1(KE$_1$)
			KEY(E_1.KE$_1$), KEY(E_2.KE$_2$),
			KEY(R.KE$_1$), KEY(W_R.LSs, W_R.KE$_2$,
			W_R.KE$_1$)
			KEYREF(R.KE$_2$ → E_2.KE$_2$)
			CHECK(W_R.KE$_2$ = E_2.KE$_2$)
			CHECK(W_R.KE$_1$ = E_1.KE$_1$)

LT		E_1(KE$_1$, R*)
		  R(KE$_2$, S_R*)
		    S_R(W_R)
		      W_R(LSs,LSe,TTs,TTe,KE$_1$,KE$_2$)
		E_2(KE$_2$)
		KEY(E_1.KE$_1$), KEY(E_2.KE$_2$),
		KEY(R.KE$_2$),
		KEY(W_R.LSs,W_R.LSe,W_R.TTs,W_R
		.KE$_1$, W_R.KE$_2$)
		KEYREF(R.KE$_2$ → E_2.KE$_2$)
		CHECK(W_R.KE$_1$ = E_1.KE$_1$)
		CHECK(W_R.KE$_2$ = E_2.KE$_2$)
			E_2(KE$_2$, R*)
			  R(KE$_1$, S_R*)
			    S_R(W_R)
			      W_R(LSs,LSe,TTs,TTe,KE$_2$,KE$_1$)
			E_1(KE$_1$)
			KEY(E_1.KE$_1$), KEY(E_2.KE$_2$),
			KEY(R.KE$_1$),
			KEY(W_R.LSs,W_R.LSe,W_R.TTs,
			W_R.KE$_2$, W_R.KE$_1$)
			KEYREF(R.KE$_1$ → E_1.KE$_1$)
			CHECK(W_R.KE$_2$ = E_2.KE$_2$)
			CHECK(W_R.KE$_1$ = E_1.KE$_1$)

VT		E_1(KE$_1$, R*)
		  R(KE$_2$, S_R*)
		    S_R(W_R)
		      W_R(VTs, VTe, KE$_1$, KE$_2$)
		E_2(KE$_2$)
		KEY(E_1.KE$_1$), KEY(E_2.KE$_2$),
		KEY(R.KE$_2$), KEY(W_R.VTs,
		W_R.KE$_1$, W_R.KE$_2$)
		KEYREF(R.KE$_2$ → E_2.KE$_2$)
		CHECK(W_R.KE$_1$ = E_1.KE$_1$)
		CHECK(W_R.KE$_2$ = E_2.KE$_2$)
			E_2(KE$_2$, R*)
			  R(KE$_1$, S_R*)
			    S_R(W_R)
			      W_R(VTs, VTe, KE$_2$, KE$_1$)
			E_1(KE$_1$)
			KEY(E_1.KE$_1$), KEY(E_2.KE$_2$),
			KEY(R.KE$_1$), KEY(W_R.LSs, W_R.KE$_2$,
			W_R.KE$_1$)
			KEYREF(R.KE$_1$ → E_1.KE$_1$)
			CHECK(W_R.KE$_2$ = E_2.KE$_2$)
			CHECK(W_R.KE$_1$ = E_1.KE$_1$)

BT		E_1(KE$_1$, R*)
		  R(KE$_2$, S_R*)
		    S_R(W_R)
		      W_R(VTs,VTe,TTs,TTe,KE$_1$,KE$_2$)
		E_2(KE$_2$)
		KEY(E_1.KE$_1$), KEY(E_2.KE$_2$),
		KEY(R.KE$_2$),
		KEY(W_R.VTs,W_R.VTe,W_R.TTs,W_R
		.KE$_1$, W_R.KE$_2$)
		KEYREF(R.KE$_2$ → E_2.KE$_2$)
		CHECK(W_R.KE$_1$ = E_1.KE$_1$)
		CHECK(W_R.KE$_2$ = E_2.KE$_2$)
			E_2(KE$_2$, R*)
			  R(KE$_1$, S_R*)
			    S_R(W_R)
			      W_R(VTs,VTe,TTs,TTe,KE$_2$,KE$_1$)
			E_1(KE$_1$)
			KEY(E_1.KE$_1$), KEY(E_2.KE$_2$),
			KEY(R.KE$_1$),
			KEY(W_R.VTs,W_R.VTe,W_R.TTs,
			W_R.KE$_2$, W_R.KE$_1$)
			KEYREF(R.KE$_1$ → E_1.KE$_1$)
			CHECK(W_R.KE$_2$ = E_2.KE$_2$)
			CHECK(W_R.KE$_1$ = E_1.KE$_1$)

*	Nest	Unable	Unable

Rule 4. Transformation of temporal attributes of relationship

If: The relationship between two entity types $E1$ and $E2$ has temporal aspect $A*$ (the attribute A is supported with temporal aspect *),
Then:

- Create the complex element S_A_R which is nested in R, and add two the constraint attributes minOccurs = "0" and maxOccurs = "unbounded".
- Create the complex element W_A_R which is nested in S_A_R and add the timestamp attributes corresponding with temporal aspect * shown in Table 3, simultaneously add the attributes corresponding with the key attributes of the entity type in relationship R
- The primary key constraint of the complex element W_A_R is created by combining the keys of the entity types in relationship R and the partial key T' of the temporal aspect * shown in Table 3.

3.2 Example

Figures 11 and 12 demonstrate an example for transformation of TimeER model into Temporal XML schema.

Apply these rules, the result of transformation is Temporal XML schema which is represented by the hierarchical model shown in Fig. 12.

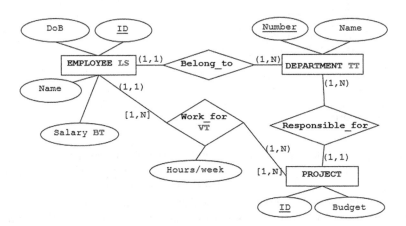

Fig. 11 The original TimeER model

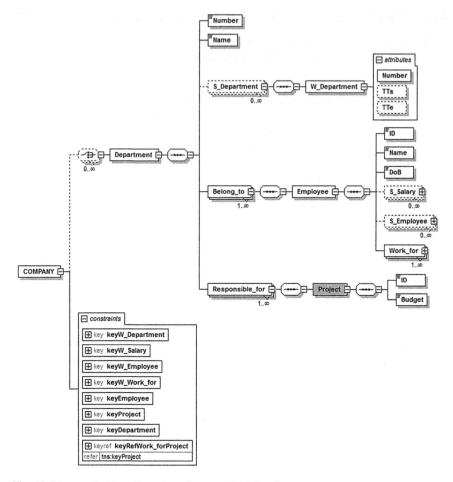

Fig. 12 The result of transformation of Temporal XML schema

4 Conclusion

In this paper, we have proposed a method of transformation the TimeER into Temporal XML schema. This proposal is an extension of the method to transform the traditional ER model into XML schema. Accordingly, the paper proposes the rules which transform the temporal components into TimeER model.

Our further research will be the standardization of the Temporal XML schemas, applying for designing the Temporal XML documents which are not redundancy data. In addition, we are also interested in researching on design of temporal ontology from the temporal conceptual databases.

References

1. Gregersen, H., Jensen, C.S.: Temporal entity-relationship models—a survey. In: IEEE Transactions on Knowledge an Data Engineering, pp. 464–497 (1999)
2. Khadija, A.: A temporal extension of XML. In: Czech Technical University in Prague, Faculty of Electrical Engineering, Department of Computer Science and Engineering (2008)
3. Kristin K.: Temporal XML. In: Paper for the Seminar in Database Systems, Universität Zürich (2010)
4. Embley, D., Mok, W.Y.: Developing XML documents with guaranteed "good": properties. In: Proceedings of the 20th International Conference on Conceptual Modeling, pp. 426–441 (2001)
5. Hoang, Q., Vo, H.L.M., Vo, V.H.: Mapping of nested multi-valued composite attributes: an addition to conceptual design for XML schemas. In: Asian Conference on Information Systems 2014—ACIS 2014, pp. 151–158 (2014)
6. Massimo F., Donatella G., Angelo M. and Carla P.: A graph-theoretic approach to map conceptual designs to xml schemas. ACM Trans. Database Syst. **38**(1), Article 6 (2013)
7. Sung J., Woohyun K.: Mapping rule for ER to XML using XML schema. In: The 2007 Southern Association for Information Systems Conference SAIS, pp. 211–216 (2007)
8. Massimo, F., Donatella, G., Angelo, M., Carla, P.: From entity relationship to XML schema: a graph-theoretic approach. Lect. Notes Comput. Sci. **5679**, 165–179 (2009)
9. Jensen, C.S.: Temporal Database Management. Dr.techn. thesis by Christian S. Jensen. http://people.cs.aau.dk/~csj/Thesis/ (2000)
10. Elmasri, R., Navathe, S.B.: Fundamentals of Database Systems, 6th edn. Addison-Wesley Publishers, United States of America (2011)
11. Chen, P.P.: The entity-relationship model—toward a unified view of data. ACM Trans. Database Syst. **1**(1), 9–36 (1970)
12. Flavio, R., Vaisman, A.A.: Temporal XML: modeling, indexing, and query processing. J. VLDB J.—Int. J. Very Large Data Bases **17**(5), 1179–1212 (2008)
13. Khadija, A., Jasoslav, P.: A comparison of XML-based Temporal Models. Adv. Internet Based Syst. Appl. LNCS **4879**, 339–350 (2009)
14. Hoang, Q., Thuong, P.: Expressing a temporal entity-relationship model as a traditional entity-relationship model. In: 7th International Conference, ICCCI 2015, Madrid, Spain, pp. 483–491, 21–23 Sept 2015
15. Hoang, Q., Nguyen, V.T.: Extraction of a temporal conceptual model from a relational database. Int. Inf. Database Syst. **6591**, 57–66 (2013)

Job Scheduling in a Computational Cluster with Multicore Processors

Tran Thi Xuan and Tien Van Do

Abstract In this paper, we investigate a job scheduling problem in a heterogeneous computing cluster built from servers with multicore processors. Dynamic Power Management technique is applied, where the delay to bring a server from the sleep to the active state is taken into account. Numerical results show that the computing resources are utilized more efficiently if multi jobs are executed in parallel. Furthermore, scheduling policy should investigate machine parameters at core-level to achieve the best efficiency.

Keywords Heterogeneous cluster · Multi-core · Dynamic Power Management · Job scheduling

1 Introduction

The explosive advance of multi-core technology in processor design opens a new era of computer science [1–4]. It is indicated that deploying multiple low-frequency processors (cores) on the same chip space enhances the overall performance while it retains the same power in comparison to a high-frequency single processor. Moreover, power management techniques can be applied at various levels as machine-, CPU-, and core-level, which brings the most efficient manner of energy savings [5–8].

Typically, present computing systems are heterogeneously built with multi-core servers. Job scheduling is a critically serious problem in such those systems [9–12]. To deal with heterogeneity, energy and performance aware scheduling policies were studied in [13] to provide either energy efficiency or high performance for systems. Buffering schemes were investigated in a heterogeneous cluster in [14]. Energy

T.T. Xuan · T.V. Do (✉)
Department of Networked Systems and Services, Budapest University of Technology
and Economics, Budapest, Hungary
e-mail: do@hit.bme.hu

© Springer International Publishing Switzerland 2016   75
T.B. Nguyen et al. (eds.), *Advanced Computational Methods*
for Knowledge Engineering, Advances in Intelligent Systems
and Computing 453, DOI 10.1007/978-3-319-38884-7_6

management and job scheduling in computing clusters with multi-core servers, however, have not been fully studied.

In this paper, we study the scheduling issue and energy consumption in a heterogeneous, multi-core cluster. Three job scheduling scenarios are investigated, where either a job can occupy the entire computing resources of a server or multiple jobs can be run concurrently on processing cores of a server. As servers are heterogeneous, previously studied Energy Efficiency (EE) and High Performance (HP) priority based scheduling policies [13, 14] are also subsequently applied and examined in the context of multi-processing. For energy savings, Dynamic Power Management technique is applied to wholly or partially switch off an idle server. A delay cost to bring a sleeping server to its active state is taken into account.

The rest of paper is organized as follows. Section 2 presents cluster model, job scheduling scenarios, and measured metrics. The numerical simulation results are presented in Sect. 3. Finally, Sect. 4 concludes the paper.

2 System Model and Scheduling Scenarios

2.1 System Model

We consider a heterogeneous computing cluster of commercial multi-core servers, of which parameters follow SPECpower_ssj2008 benchmark of the Standard Performance Evaluation Corporation (SPEC) [15]. The cluster (illustrated in Fig. 1) includes

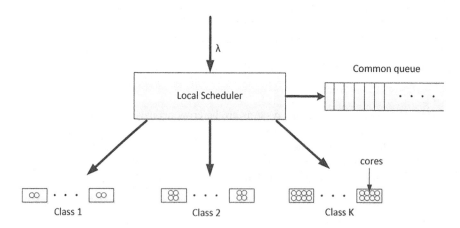

Fig. 1 A common queue cluster of multi-core servers

- K classes of identical servers, therein class i ($i = 1, ..., K$) consists of M_i servers;
- a common buffer to store waiting jobs when none server is available as proposed in [14];
- and a local scheduler, which is aware of servers' performance and power consumption.

Let S denote set of server types, then $|S| \leq K$. The considered parameters of server type s ($s \in S$) are:

- N_s (the number of homogeneous cores),
- C_s (the ssj_ops value, defined by the number of operations per second),
- $P_{ac,s}$ (the average active power),
- and $P_{id,s}$ (power consumption when server is active idle).

We assume that the considered system is capable for executing jobs that:

- have service demand unknown to the local scheduler,
- can be executed on any server,
- are attended to scheduler by the First Come First Served (FCFS) service policy, and
- are uninterrupted during being executed (non-preemptible property).

We apply Dynamic Power Management (DPM) technique for energy efficiency. In a previous study [16], authors addressed the issues of DPM and indicated that a server should wait a certain period before going to sleep state. Authors in [17] proposed SoftReactive policy where each server will independently decide its optimal waiting time in idle state (t_{wait}) before switching off. The optimal waiting time must ensure the equalization between waiting energy consumption ($t_{wait} * P_{idle}$) and energy consumption in setup mode ($T_{setup} * P_{max}$). Thus, it can be calculated by:

$$t_{wait} = \frac{P_{max}}{P_{idle}} * T_{setup} \qquad (1)$$

If new job arrives in t_{wait}, server will be back to active immediately without any cost. Otherwise, it will be set to sleep state with $P_{sleep} = 0W$. In our study, we exploit the best sleep state with the pair of $P_{sleep} = 0W$ and $T_{setup} = 20s$, as indicated [16].

2.2 Job Distribution Scenarios

In this section, we investigate three scenarios to distribute jobs into the considered cluster, where a single job or multiple jobs can be concurrently executed in a server. We assume the symmetric multiprocessing (SMP) technology is deployed in scenarios of multiple jobs; i.e. when a new job comes, job will be served immediately if there exists unused core in system. Three scenarios are described as follows.

- *Single-job scenario*: At one time, a server of type s ($s \in S$) is capable to handle a single job at full workload capacity C_s and consume its maximum active power $P_{ac,s}$. DPM is deployed at machine-level to switch off one server of type s after it waits $t_{wait,s}$ seconds on idle state with no new incoming job.
- *Full-active scenario*: each core of a server can be considered as an individual computing element to handle a job. The computing capacity of cores is assumed as equal. That means a server type s enables N_s jobs to be simultaneously executed with the performance capacity of C_s/N_s. The DPM technique is applied at machine level as same as Single-job scenario. Server stays fully active and consumes the maximum active power $P_{ac,s}$ while jobs are processed.
- *Dynamic-active scenario*: The scheduling policy is as same as Full-active scenario; N_s jobs can be executed in parallel in one server type s and each core handles one job with the performance capacity of C_s/N_s. The DPM dynamically switches off idle server as well as unused cores for energy savings. That means when k cores execute jobs concurrently in a server type s, the server performs at the capacity of C_s^k and consumes power $P_{ac,s}^k$ as the following formulas (according to the linear model of energy consumption in [18]):

$$P_{ac,s}^k = P_{id,s} + \frac{k}{N_s} * (P_{ac,s} - P_{id,s}), s = 1, ..., K \tag{2}$$

$$C_s^k = \frac{k}{N_s} * C_s, s = 1, ..., K \tag{3}$$

We should note that time to switch on cores from sleep state is too negligible to impact on system.

After applying EE or HP policy, routing process is implemented as illustrated in Algorithm 1.

Algorithm 1 Schedule

---

  for $i = 1 \rightarrow K$ **do**
    for $j = 1 \rightarrow M_i$ **do**
      if server (i,j) is FREE **then**
        if busy_cores of (i,j) are MAXIMUM **then**
          free_server $\leftarrow (i,j)$
          GOTO ALLOCATE
        end if
      end if
    end for
  end for
  ALLOCATE:
  if found *free_server* **then**
    ROUTE job to *free_server*
  else
    ROUTE job to Common Queue
  end if

Table 1 Notations

K	Number of server classes
M_i	Number of servers in class i
λ	System arrival rate
μ_i	Service rate of each server in class i
μ_{system}	Service rate of entire system
U (ρ_{system})	Average system utilization
WT	The average waiting time per job
RT	The average response time per job
AIE	The average idle energy consumption per job
AOE	The average operating energy consumption per job
PJ	The average number of processed jobs per second

2.3 Notations

All notations of system parameters and measurement metrics are listed in Table 1.

3 Numerical Results

We build a simulation software to implement our presented study approach. The software is run with the stop condition of ten millions of completions. For reliability, simulation is run with the confident level of 99 %. The accuracy (i.e. the ratio of the half- width of the confidence interval and the mean of collected observations) is less than 0.01 for all measured metrics. Computing servers and system loads are described in Sect. 3.1. Numerical results are presented in Sect. 3.2.

3.1 Server Specifications

The studied cluster model has three server classes ($K = 3$), of which each consists of eight identical servers ($M_i = 8, i = 1, ..., K$). Chosen server types and their parameters are presented in Table 2. It is noted that computing capacity of server type i, C_i, presented (ssj_ops) and its power consumption, $P_{ac,i}$, are measured at 100 % target load. The optimal waiting time $t_{wait,i}$ is also calculated according to Eq. 1 and shown in Table 2.

Table 2 Server parameters

Server type	N_*	C_*	$P_{ac,*}$ (W)	$C_*/P_{ac,*}$	$P_{id,*}$ (W)	$t_{wait,*}$ (s)
Fujitsu TX100 S3p (Intel Xeon E3-1240V2) [19]	4	467481	72.5	6450	14.1	102.8
Acer AR380 F2(Intel Xeon E5-2640) [20]	12	990555	254	3904	81.9	62.0
Acer AR380 F2(Intel Xeon E5-2665) [21]	16	1347230	409	3298	103	79.4

Table 3 Ranks of servers

Server type	HP based rank	EE based rank
Intel Xeon E3-1240V2	$r_p(1) \approx 0.35$	$r_e(1) = 1.0$
Intel Xeon E5-2640	$r_p(2) \approx 0.74$	$r_e(2) \approx 0.61$
Intel Xeon E5-2665	$r_p(3) = 1.0$	$r_e(3) \approx 0.51$

Servers are classified according to the ranking functions in [14] as shown in Table 3. Based on the chosen priority, servers in class with higher rank are prior.

We assume that the inter-arrive time and the execution time of jobs are exponentially distributed with means of $1/\lambda$ and $1/\mu$, respectively. Let μ_1, μ_2 and μ_3 denote the service rates, if jobs are scheduled to a server of type Intel Xeon E3-1240V2, Intel Xeon E5-2640, and Intel Xeon E5-2665, respectively. Service rate of the whole system can be calculated as

$$\mu = \sum_{i=1}^{K} M_i \times \mu_i, \tag{4}$$

and traffic volume carried by system is defined as

$$\rho = \frac{\lambda}{\mu}. \tag{5}$$

If we assume that each job requires a computing capacity equivalent to 1347230 (ssj_ops) in average, which means the average execution time of jobs is one second if jobs are routed to an Intel Xeon E5-2665 server. According to priority classification, we achieve $\mu_1 = 0.35/s$ $\mu_2 = 0.74/s$, and $\mu_3 = 1/s$. Hence, the total service rate of system equals to $8 * (0.35 + 0.74 + 1) = 16,72$. We apply the system traffic volumes equal to 0.20, 0.30, 0.40, 0.50, and 0.60 in our simulation runs. It is also worth to note that in multi-job scenarios, without losing generality, every core of machine type i has the same ssj_ops value. That means each core handles jobs with service rate μ_i/N_i, see Table 4.

Table 4 Capacity per core

Server type	$C_*/core$	$\mu_*/core$
Intel Xeon E3-1240V2	116870	$\approx 1.7e-3$
Intel Xeon E5-2640	82546	$\approx 1.2e-3$
Intel Xeon E5-2665	84202	$\approx 1.22e-3$

3.2 Obtained Results

We present the system performance in term of mean response time per job, mean waiting time per job, and average number of processed jobs per second when EE and HP policy are subsequently applied.

Figure 2 shows the average response time regarding to scheduling scenarios (single-job, full-active, and dynamic-active), when the system changes from low to high loads. It is obvious that when a job occupies the whole computing resources of a server (single-job scenario), it is processed fastest. However the average number of processed jobs per second indicates that scheduling scenarios do not impact the overall performance of system (see Fig. 3). Figure 4 shows that job have to wait longest in single job scenario due to the lack of available computing resource in system. Furthermore, the HP scheduling policy does not achieve the advantage of high performance if it is considered at machine-level while the computation is performed at core-level in multi-job scenarios.

Figure 5 plots the average of energy consumption in operation (i.e. the sum of energy consumption in both idle state and processing state) when the two policies (EE and HP) are alternatively applied.

The result shows that keeping a server fully active when its utilization is less (in Full-active scenario) causes a energy waste, while dynamic-active scheduling can yield significant savings. It is also observed that server-level applied EE policy may loose the advantage in multi-job scenarios.

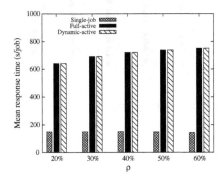 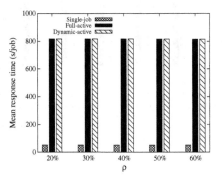

Fig. 2 Mean response time per job versus system load

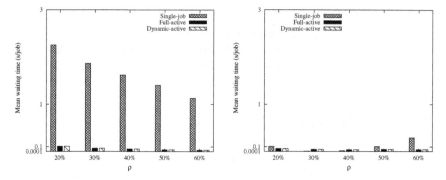

Fig. 3 Mean waiting time per job versus system load

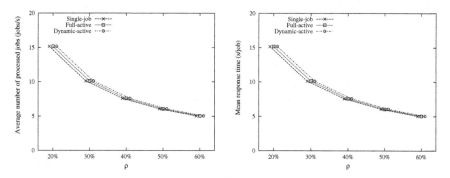

Fig. 4 Average number of processed job per second versus system load

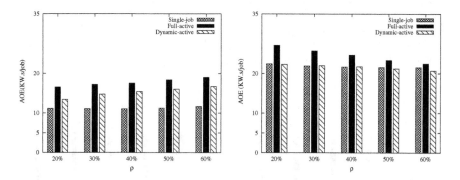

Fig. 5 Average operation energy consumption per job versus system load

4   Conclusion

In this paper, we investigate various scheduling scenarios in a heterogeneous computational cluster of multi-core servers. The Dynamic Power Management technique is appropriately deployed with setup cost at server-level or even at core-level for the purpose of energy savings. Energy Efficiency (EE) and High Performance (HP) policies are also applied for evaluation.

The numerical results show that as multi-processing is applied in multi-core system, the computing resources are utilized efficiently with the same overall performance. It is observed that the scheduling policies may loose their properties in the multi-processing context if the scheduler is aware of machine characteristics at server level. Therefore, a core-level awareness may play a key role for job scheduling in multi-core computing system.

References

1. Roy, A., Xu, J., Chowdhury, M.: Multi-core processors: A new way forward and challenges. In: International Conference on Microelectronics, 2008. ICM 2008, pp. 454–457 (2008)
2. Wang, L., Tao, J., von Laszewski, G.: Multicores in cloud computing: Research challenges for applications. J. Comput. 5(6) (2010)
3. Chapman, M.T.: The benefits of dual-core processors in high-performance computing (2005)
4. NVIDIA: The benefits of multiple cpu cores in mobile devices (2011)
5. Grochowski, E., Ronen, R., Shen, J., Wang, P.: Best of both latency and throughput. In: Proceedings of IEEE International Conference on Computer Design: VLSI in Computers and Processors, 2004. ICCD 2004, pp. 236–243 (2004)
6. NVIDIA: Variable smp—a multi-core cpu architecture for low power and high performance, Technical Report, NVIDIA's Project Kal-El, NVIDIA Corporation (2012)
7. Kolpe, T., Zhai, A., Sapatnekar, S.: Enabling improved power management in multicore processors through clustered dvfs. In: Design. Automation Test in Europe Conference Exhibition (DATE) 2011, pp. 1–6 (2011)
8. Qi, X., Zhu, D.-K.: Energy efficient block-partitioned multicore processors for parallel applications. J. Comput. Sci. Technol. 26(3), 418–433 (2011)
9. Shieh, W.-Y., Pong, C.-C.: Energy and transition-aware runtime task scheduling for multicore processors. J. Parallel Distrib. Comput. 73, 1225–1238 (2013)
10. Chai, L., Gao, Q., Panda, D.: Understanding the impact of multi-core architecture in cluster computing: a case study with intel dual-core system. In: Seventh IEEE International Symposium on Cluster Computing and the Grid, 2007. CCGRID 2007, pp. 471–478 (2007)
11. Nesmachnow, S., Dorronsoro, B., Pecero, J., Bouvry, P.: Energy-aware scheduling on multicore heterogeneous grid computing systems. J. Grid Comput. pp. 1–28 (2013)
12. Papazachos, Z.C., Karatza, H.D.: Gang scheduling in multi-core clusters implementing migrations. Future Gener. Comput. Syst. 27, 1153–1165 (2011)
13. Zikos, S., Karatza, H.D.: Performance and energy aware cluster-level scheduling of compute-intensive jobs with unknown service times. Simul. Model. Pract. Theory 19(1), 239–250 (2011)
14. Do, T.V., Vu, B.T., Tran, X.T., Nguyen, A.P.: A generalized model for investigating scheduling schemes in computational clusters. Simul. Model. Pract. Theory 37, 30–42 (2013)
15. Standard Performance Evaluation Corporation. http://www.spec.org/
16. Gandhi, A., Harchol-Balter, M., Kozuch, M.A.: The case for sleep states in servers. In: Proceedings of the 4th Workshop on Power-Aware Computing and Systems, HotPower'11, pp. 2:1–2:5, ACM, New York, NY, USA (2011)

17. Gandhi, A., Harchol-Balter, M., Kozuch, M.A.: Are sleep states effective in data centers? In: Proceedings of the 2012 International Green Computing Conference (IGCC), IGCC'12, pp. 1–10, IEEE Computer Society, Washington, DC, USA (2012)
18. Ellision, B., Minas, L.: Energy Efficiency for Information Technology: How to Reduce Power Consumption in Servers and Data Centers. Intel press (2009)
19. SPEC: Fujitsu primergy tx100 s3p (intel xeon e3-1240v2) machine (2012). http://www.spec.org/power_ssj2008/results/res2012q3/power_ssj2008-20120726-00519.html
20. SPEC: Acer Incorporated Acer ar380 f2 (intel xeon e5-2640) machine (2012). http://www.spec.org/power_ssj2008/results/res2012q3/power_ssj2008-20120525-00481.html
21. SPEC: Acer Incorporated Acer ar380 f2 (intel xeon e5-2665) machine (2012). http://www.spec.org/power_ssj2008/results/res2012q3/power_ssj2008-20120525-00479.html

Modeling Multidimensional Data Cubes Based on MDA (Model-Driven Architecture)

Truong Dinh Huy, Nguyen Thanh Binh and Ngo Sy Ngoc

Abstract As data warehouse (DWH) expands its scope increasingly in more and more areas, its application development processes still have to face significant challenges in semantically and systematically integrating heterogeneous sources into data warehouse. In that context, we propose an ontology-based multi dimensional data model, aiming at populating data warehousing systems with reliable, timely, and accurate information. Furthermore, in our approach, MDA (Model-Driven Architecture) principles and AndroMDA system are utilized to specify as well as to generate source code from UML (Unified Modeling Language) class diagrams, which are used to model the ontology-based multi dimensional data model in the context of object oriented concepts. As a result, this approach enables interoperability among a class of DWH applications, which are designed and developed based on our proposed solution.

Keywords API (Application Programming Interface) · Data Warehouses (DWH) · MDA (Model-Driven Architecture) · Andromda · OLAP (Online Analytical Processing) · UML (Unified Modeling Language)

T.D. Huy (✉)
Duy Tan University, Da Nang, Vietnam
e-mail: huy.truongdinh@gmail.com

N.T. Binh
International Institute for Applied Systems Analysis (IIASA),
Schlossplatz 1, 2361 Laxenburg, Austria
e-mail: nguyenb@iiasa.ac.at

N.S. Ngoc
Hue University, Hue, Vietnam
e-mail: ngongoc11906@gmail.com

© Springer International Publishing Switzerland 2016   85
T.B. Nguyen et al. (eds.), *Advanced Computational Methods for Knowledge Engineering*, Advances in Intelligent Systems and Computing 453, DOI 10.1007/978-3-319-38884-7_7

1 Introduction

MDA is a software design approach to develop information systems [1, 2]. It provides a set of guidelines for the structuring of specifications, which are expressed as models. MDA is a kind of domain engineering and supports model-driven engineering of software systems [2]. At the heart of the MDA approach are models which represent the information system developed while the implementation in a certain technology is obtained through model transformation [1–3]. From the benefits of MDA: productivity, interoperability, portability, etc., especially the modeling, many organizations are now looking at the ideas of MDA as a way to organize and manage their application solutions. Tool vendors are explicitly referring to their capabilities in terms of "MDA compliance", and the MDA lexicon of platform-specific and platform-independent models is now widely referenced in the industry [1, 2]. Besides, many researchers also have examined many of their works on MDA [1, 3].

DWH and OLAP are essential elements of decision support [4–10]. OLAP systems organize data using the multidimensional paradigm in the form of data cubes, each of which is a combination of multiple dimensions with multiple levels per dimension [8, 11].

The main contribution of this paper is to propose a semantic ontology-based model approach for multi dimensional databases, which is the fundamental part of in DWH system. In the proposed approach, a combined use of both semantics in terms of ontology—based multidimensional data model up to MDA-based framework will be introduced, providing well-defined integrated, semantic foundation for interoperability of various aspects of the DWH application development processes. In this context, the AndroMDA software, which is an instance of the MDA-based concept, is used to generate source code from class diagrams of process modeling. Moreover, the integrated use of MDA and ontology provides abilities to develop a a class of data warehousing applications in a systematic and sematic manner. Thus, the well-defined descriptions of schema-based and content-based semantics in ontology can facilitate the truly interoperability in DWH environments.

The remainder of this paper is organized as follows: in Sect. 2, we discuss about related works. Then in Sect. 3, we introduce the ontology-based multidimensional conceptual model and some related theories. In Sect. 4, we present how to modeling multidimensional data cubes based on MDA, how to use AndroMDA to generate source code and architecture of MDA-based and Ontology-based framework. The paper concludes with Sect. 5, which presents our current and future works.

2 Related Work

There is some research works on multi-dimensional data model is mentioned in [10, 12–17]. In particular, in our previous research works have proposed a suitable mutidimensional data model for OLAP. Besides, we have also proposed the very elegant manners of definitions of three cube operators, namely *jumping*, *rollingUp* and *drillingDown*, the modeling of the conceptual multidimensional data model in term of classes by using UML [8, 9, 18], which is a key in the MDA approach [4–6]. In MDA models are more than abstract descriptions of systems as they are used for model transformations and code generation—they are the key part defining a software system [2, 3, 19, 20].

Currently, there are some software platforms which enable software architects and developers to build and integrate software applications in an automated and industrialized way, such as: OpenMDX, Rational XDE (Rational Extended Development Environment), etc. [1–3]. In particular, recently AndroMDA is an open source of MDA framework [21]—it takes any number of models (usually UML models stored in XMI produced from case-tools) combined with any number of AndroMDA plugins (cartridge and translation-libraries) and produces any number of custom components. It can generate components for any language: Java, .Net, HTML, PHP, etc., that we just write (or customize existing) plugins to support it [2, 20].

Meanwhile, various researches have been done to bridge the gap between the metamodel-based integration and semantic technology, supporting the business model interoperability [5, 10, 22, 23]. To the best of our knowledge, in the state-of-the-art research and practice, little related research on the argument in this paper is found. From our point of view, these proposals are oriented to the conceptual and logical design of the data warehouses development, and do not seriously taking into account important aspects such as the challenges of software devlopment processes with structural and semantic heterogeneity, the main focus of this research.

In our work, using AndroMDA means one main thing: write less code. Not only that, AndroMDA also lets us create better applications and maintain better order on large projects. With AndroMDA, classes diagram become a living part of our application. They are used to generate large portions of our application, and hence always reflect the current state of the system. When we need to modify application, we change the model first, regenerate the code, and then add or update custom code as necessary. We get a production quality application out of assets that we had to create anyway.

3   Multi Dimensional Conceptual Model

In our previous work [5, 6, 8], we have introduced hierarchical relationships among dimension members by means of one hierarchical domain per dimension. A hierarchical domain is a set of dimension members, organized in hierarchy of levels, corresponding to different levels of granularity. It allows us to consider a dimension schema as a partially ordered set of levels. In this concept, a hierarchy is a path along the dimension schema, beginning at the root level and ending at a leaf level. Moreover, the recursive definitions of two dimension operators, namely *ancestor* and *descendant*, provide abilities to navigate along a dimension structure. In a consequence, dimensions with any complexity in their structures can be captured with this data model.

In this section, we present briefly the concepts and basic ideas of proposed multi dimensional data model based on ontology, especially the ontology-based descriptions of dimensions, dimension hierarchies, facts, measures and cubes in data warehouse. Then the ontology-driven framework can solve main structural and semantic problems at graph-based components in term of illustrated examples.

3.1 Motivating Example

The GAINS data warehouse [9–11, 18] from hundreds of *regions* is presented as a running example. The calculated *Emission* data are stored in the warehouse in a large table, called fact table. The warehouse contains a set of **dimensions** (*Fuel Activity*, *Sector*, *Region*, *Scenario*, *Pollutant* and *Time*) describing information related to the fact table. The **dimension schemas** with hierarchical **levels** of these dimensions are described in Fig. 1.

In addition, for this warehouse, a possible fact includes calcualted *Emission* data. This fact can be analyzed with respect to a given *pollutant*, a GAINS *region*, and a selected *year* within a given GAINS *scenario*. Thus, a fact table *Emissions* can be illustrated in following table Table 1.

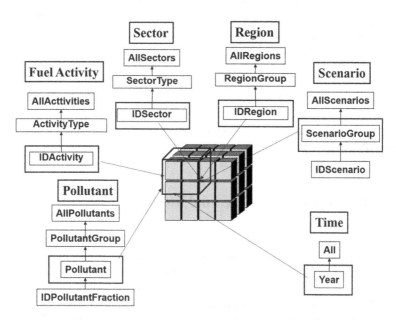

Fig. 1 Examples of dimension schemas

Table 1 The *Emissions* fact table

SECTOR	ACT_CATEGORY	IDREGIONS	IDSCENARIOS	POLLUTANT	YEAR	EMISS
CON	BIOMASS	AUST_WHOL	CP_WEO_2011	SO2	1995	0
RES_COM	BIOMASS	AUST_WHOL	CP_WEO_2011	SO2	1990	2.054
CON	LIQUID	AUST_WHOL	CP_WEO_2011	SO2	2015	2.4501418256824771355410558678227 1395956
RES_COM	COAL	AUST_WHOL	CP_WEO_2011	SO2	2030	0.266839981784736286664719920712 3483410114
CON	COAL	AUST_WHOL	CP_WEO_2011	SO2	2000	0.249258848723761611822827223563 8403550803
POWER_PL	BIOMASS	AUST_WHOL	CP_WEO_2011	SO2	2000	0.86737225297188274533
IND_COMB	GAS	AUST_WHOL	CP_WEO_2011	SO2	2020	0
POWER_PL	COAL	AUST_WHOL	CP_WEO_2011	SO2	2035	0.881379412915446175463288156932 0989364371
IND_COMB	LIQUID	AUST_WHOL	CP_WEO_2011	SO2	2000	3.144654973769738783456725984400 87960522
NONEN	LIQUID	AUST_WHOL	CP_WEO_2011	SO2	2025	0
RES_COM	LIQUID	AUST_WHOL	CP_WEO_2011	SO2	2030	1.286378482623311458302423278624 11882169
RES_COM	LIQUID	AUST_WHOL	CP_WEO_2011	SO2	1990	15.527796787347669835244815579453 36332571
POWER_PL	OTH_RENEW	AUST_WHOL	CP_WEO_2011	SO2	2010	0

3.2 Multidimensional Model

Multi dimensional data model is defined as $MD = \ < \{DC_x\}, \{FC\} >$

- DC_x: is a set of concepts defining a dimension and its related concepts in $\mathcal{D} = \{D_1, \ldots, D_x\}, x \in N$.
- FC_y: is a set of concepts each of which defines a fact in $\mathcal{F} = \{F_1, \ldots, F_y\}, y \in N$.

3.2.1 The Modeling of Dimensions

DimensionMember. This class describes the hierarchical domain of a dimension, consists of a finite set of dimension members, organized in hierarchy of levels, corresponding to different levels of granularity.

description—value of data member in dimension. Figure 2 show an example of the fuel activity dimension, which is developed in NEO4 J [24] and stored in term of graph-based data.

parent (higher-level) and *child* (lower-level)—provide hierarchical relations between dimension members. For example *COAL* is parent of *HC2, HC3, BC1, BC2* (Fig. 2).

DimensionLevel. This class describes a detailed level of a dimension denoted as

$$DLC_x = \ < Lname, \ dom(DLC_x) >, \text{ where:}$$

Lname—name of a level, e.g. *IDActivity, Activity Type*
ancestor (higher-level) and *descendant* (lower-level)—provide the descriptions of ordered relation in level of detail.

hasDimensionMembers (symmetric property)—provide the relationship between level and its corresponding dimension members $(dom(DLC_x))$.

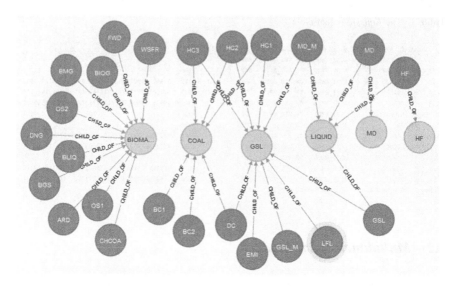

Fig. 2 An example of the graph-based *fuel activity* dimension

RootLevel. This class defines *Root* level of a dimension:

$$DRLC_x = \ <All, dom(DRLC_x)>$$

- All—name of a root level, e.g. *AllActivities*
- $dom(DRLC_x)$ has one and only one instance
- The value of ancestor is NULL.

LeafLevel. This class describes Leaf level of a dimension:

$$DLLC_x = \ <Leafname, dom(DLLC_x)>$$

- *Leafname*—name of a leaf level. For example, *IDActvity* is a leaf level.
- The value of descendant is NULL in this case.

DimensionHierarchy. This class describes a hierarchy of detailed levels of a dimension:

$$DHC_x = \ <Hname, DRLC_x, \{DLC_x\}, DLLC_x>$$

- *Hname*—name of a hierarchy
- ContainLevels (symmetric property)—provide the relationship between hierarchy and its corresponding level $(DRLC_x, \{DLC_x\}, DLLC_x)$

In this context, the *fuel activity* dimension is hierachically organized as follows *AllActivities- > ActType- > IDActivity*

DimensionSchema. This class describes a dimension schema:

$$\text{DSC}_x = <DSname, \{\text{DHC}_x\}>$$

- DSname—name of the dimension's schema
- hasHierarchies (symmetric property)—provide the relationships between schemas and corresponding dimension hierarchies ($\{\text{DHC}_x\}$)

Dimension. This class describes a dimension as

$$\text{DC}_x = <Dname, \text{DSC}_x, \text{dom}(\text{DC}_x)>$$

- *Dname*—name of a dimension
- *hasSchema* (symmetric property)—provide the relationship between dimension and its corresponding schema (DSC_x)
- *hasLevels* (symmetric property)—provide the relationship between dimension and its corresponding levels.
- dom $(\text{DC}_x) = \bigcup_{\forall l \in DC_x} dom(l))$

3.2.2 The Modeling of Facts

Fact. This class describes a fact, reflects information that have to be analyzed

$$\text{FC}_y = <\text{Fname}, \text{dom}(\text{FC}_y)>:$$

- *Fname*—name of a fact
- *hasValue*—specify a numerical domain where a fact value is defined dom $((\text{FC}_y))$

3.2.3 Dimension Constraints

Each multidimensional data model covers a number of dimensions, among which, there could be a dimension member of a dimension x may be may be associated with a number of different dimension members of a dimension y. Hence, in such data model, dimension constraints are specified by x-y combinations. For example, as illustrated in Fig. 3, in the sector 'industrial boilers'(*IN_BO*) the associated fuel activities are the various fuels that are used in industrial boilers, i.e., *COAL*, *OIL*, etc.

Fuel activities may be further subdivided, e.g., hard coal grade 1 (*BC1*), hard coal grade 2 (*BC2*). Some case studies about dimension constraints have been introduced in [11].

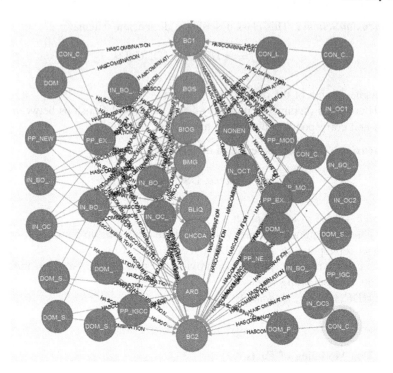

Fig. 3 An example of the graph-based *fuel activity—sector* combinations

4 Modeling Multidimensional Data Cubes Based on MDA

We extend *megamodel* in [19] later to illustrate the relationships between MDE concepts and ontology technologies. Afterwards, AndroMDA is proposed to generate source code from the UML class diagrams. AndroMDA generate entity classes, data access objects, and service interfaces and base classes. In order to have a common framework and solutions based on AndroMDA, ontology-based models has been taken into consideration in modeling, creating tool(s) and presented in session 3. In this context, concepts and ontologies will be used to generate the API, so that enables software modeler easily in specifying problem, reusing the linked design patterns. Therefore, it will help to standardize and speed up modeling as well as interoperability among DW and OLAP applications.

4.1 AndroMDA Concepts

MDA is an approach to software development based on modelling and automated transformation of models to implementations, starting at a computation independent model (CIM) towards a platform independent model (PIM) and refining those to platform specific models (PSM) from which the source codes can be generated [2, 20]. UML is the standard for developing industrial software projects and a key in the MDA approach [1–3]. MDA models are more than abstract descriptions of systems as they are used for model transformations and code generation—they are the key part defining a software system [1, 19].

In this context, AndroMDA is an open source MDA Framework, which takes UML models stored in XMI produced from case-tools combined with any number of AndroMDA plugins (cartridge and translation-libraries) and produces any number of custom components. We can generate components for any language: Java, .Net, HTML, PHP, we just write (or customize existing) plugins to support it and you're good to go [20]. Figure 4 is the diagram maps various application layers to Java technologies supported by AndroMDA.

- **Presentation Layer**: AndroMDA currently offers two technology options to build web based presentation layers: Struts and JSF. It accepts UML activity diagrams as input to specify page flows and generates Web components that conform to the Struts or JSF frameworks [2, 20].
- **Business Layer**: The business layer generated by AndroMDA consists primarily of services that are configured using the Spring Framework. These services are implemented manually in AndroMDA-generated blank methods, where business logic can be defined. These generated services can optionally be front-ended with EJBs, in which case the services must be deployed in an EJB

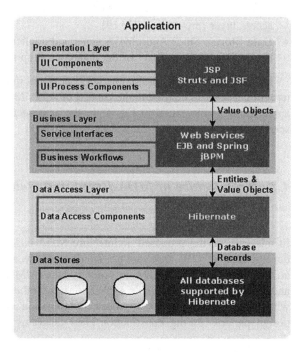

Fig. 4 Diagram maps various application layers to java technologies supported by AndroMDA

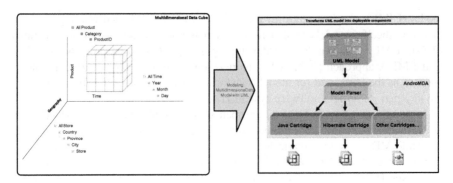

Fig. 5 Overview of the use of AndroMDA in the proposed system

container (e.g.: Jobs). Services can also be exposed as Web Services, providing a platform independent way for clients to access their functionality. AndroMDA can even generate business processes and workflows for the jBPM workflow engine [2].

Data Access Layer: AndroMDA leverages the popular object-relational mapping tool called Hibernate to generate the data access layer for applications. AndroMDA does this by generating data access objects (DAOs) for entities defined in the UML model. These data access objects use the Hibernate API to convert database records

- into objects and vice versa. AndroMDA also supports EJB3/Seam for data access layer (pre-release) [2, 19].
- **Data Stores**: Since AndroMDA generated applications use Hibernate to access the data, you can use any of the databases supported by Hibernate [19].

The modeling of the ontology-based conceptual multidimensional data model in term of classes will be used in the context of AndroMDA to generate a class of DW applications as illustrated in Fig. 5. In the next section, three main steps will be specified.

4.2 Architecture of MDA-Based and Ontology-Based Framework

In this section, the proposed architecture MDA-based approach is applied in the development of DW applications, interoperability among DW applications. Our architecture uses AndroMDA in order to takes UML models from CASE-tool(s) and generates fully deployable applications and other components that manipulate with DW. With the MDA approach, combination with capable of applying the design patterns during construction class diagrams will reduce complexity, and

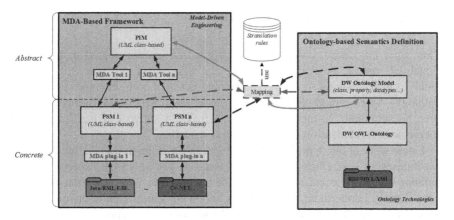

Fig. 6 Architecture of MDA-based and ontology-based framework

improve reusability. Besides, this architecture also allows the transformation of Model-Driven Engineering (MDE) models and models in ontology technologies with the aim of increasing the interoperability and integration of data in applications DW, Fig. 6 is overview of our propose architecture.

Modeling dimensions, measures and data cubes in context of an object oriented data model we used UML. All conceptual components are mapped as classes, that we have proposed in [10], this class diagram is PIM in our architecture.

According to the our architecture, an comprehensive PIM allows modelers to target multiple PSMs from a single PIM and to synchronize PIM changes into the PSM on demand and vice versa. Then, PSM generate code for several platforms and technologies: Microsoft.Net 2.0 application, Java/RMI, EJB…with the help of plug-in cartridges. The mapping between the model PIM, PSM in MDA and Ontology model using transformation rules.

Our architecture in accordance with the stages of the MDA software life cycle and take advantage of MDA approach is increased productivity, MDA can generate a large of code from models, so it saves time and delivers software solutions faster. Besides, MDA can incorporate the design patterns, templates, etc. in order to produce higher quality code and best practices of leading software experts [3].

This proposed architecture use existing tools MDA: OpenMDX, Rational XDE…in order to create applications from models. Specifically, we use tool AndroMDA—an open source code generation framework that follows the MDA paradigm. AndroMDA takes model(s) from CASE-tool(s) and generates fully deployable applications and other components [16]. AndroMDA is a great choice when we are starting a new project, and we want to save time by generating as much code as possible, besides we are building an application that stores its data in a database [16].

5 Conclusion and Future Works

This paper introduced, described and proposed an ontology-based multi dimensional data model, aiming at populating data warehousing systems with reliable, timely, and accurate information. The combined use of both semantics in terms of ontology—based multidimensional data model up to MDA-based framework has been be introduced. In this context, the AndroMDA software is used to generate source code from class diagrams of process modeling.

Future work of our approach could then be able to support users in building data warehouses in cost efficient and elastic manner that spans all aspects of cube building lifecycle, i.e. cube definition, cube computation, cube evolution as well as cube sharing. Furthermore, we will focus on the implementation of the user-driven approach among linked data cubes to make use of our concepts.

References

1. Brown, A.W.: Model Driven Architecture: Principles and Practice. Springer, New York (2004)
2. https://en.wikipedia.org/wiki/Model-driven_architecture
3. Gjoni, O.: Bizagi Process Management Suite as an Application of the Model Driven Architecture Approach for Developing Information Systems. MCSER Publishing (2014)
4. Hoang, A.-D.T., Nguyen, T.B.: An integrated use of CWM and ontological modeling approaches towards ETL processes. IEEE Computer Society (2008)
5. Hoang, A.D.T., Nguyen, T.B.: A semantic approach towards CWM-based ETL processes. In: Proceedings of the Innovations Conference for Knowledge Management, New Media Technology and Semantic Systems, 3–5 Sep 2008, Graz, Austria
6. Hoang, A.D.T., Nguyen, T.B., Tran, H., Tjoa, A.M.: Towards the Development of Large-scale Data warehouse Application Frameworks. In: Moeller, C., Chaudhry, S. (eds.) Re-conceptualizing Enterprise Information Systems, pp. 92–104. Springer, New York, USA (2012). scopusID = 2-s2.0-84859065669. ISSN: 1865-1348
7. Hoang, A.D.T., Ngo, N.S., Nguyen, T.B.: Collective cubing platform towards definition and analysis of warehouse cubes. In: Computational Collective Intelligence. Technologies and Applications—4th International Conference, ICCCI 2012. Lecture Notes in Computer Science, vol. 7654. Springer (2012). ISBN: 978-3-642-34706-1. Scopus: 2-s2.0-84870938477
8. Nguyen, T.B., Tjoa, A.M., Wagner, R.R.: An object oriented multidimensional data model for OLAP. Springer, Berlin (2000)
9. Nguyen, T.B., Wagner, F., Schoepp, W.: Cloud intelligent services for calculating emissions and costs of air pollutants and greenhouse gases. In: Intelligent Information And Database Systems (ACIIDS 2011). Lecture Notes in Computer Science (LNCS), vol. 6591. Springer (2011). ISBN 978-3-642-20038. Scopus: 2-s2.0-84872157447
10. Nguyen, T.B., Ngo, N.S.: Semantic cubing platform enabling interoperability analysis among cloud-based linked data cubes. In: CONFENIS 2014. ACM. Hanoi, Vietnam (2014). ISBN: 978-1-4503-3001-5
11. Nguyen, T.B.: Integrated assessment model on global-scale emissions of air pollutants. Studies in Computational Intelligence (2015). ISSN: 1387-3326
12. Papastefanatos, G., Petrou, I.: Publishing statistical data as linked data—the RDF data cube vocabulary. Institute for the Management of Information Systems (2013)

13. Helmich, J., Klímek, J., Nečaský, M.: Visualizing RDF data cubes using the linked data visualization model. LOD2 project SVV-2014-260100 (2014)
14. Han, J.: Data Mining: Concepts and Techniques. Elsevier Inc., (2006)
15. Stolba, N., Banek, M., Tjoa, A.M.: The security issue of federated data warehouses in the area of evidence-based medicine. IEEE (2006)
16. http://www.w3.org/TR/vocab-data-cube/
17. http://aksw.org/Projects/CubeViz.html
18. Nguyen, T.B., Schoepp, W., Wagner, F.: GAINS-BI: business intelligent approach for greenhouse gas and air pollution interactions and synergies information system. In: Proceedings of the 10th International Conference on Information Integration and Web-Based Applications & Services (iiWAS2008), Linz, Austria, 24–26 Nov 2008. ACM 2008. scopusID = 2-s2.0-70349132550. ISBN: 9781605583495
19. Parreiras, F.S.: Semantic Web and Model-Driven Engineering. IEEE (2012)
20. https://en.wikipedia.org/wiki/OpenMDX
21. http://andromda.sourceforge.net/andromda-documentation/getting-started-java
22. Höfferer, P.: Achieving business process model interoperability using metamodels and ontologies. In: Proceedings of the 15th European Conference on Information Systems (ECIS2007), Switzerland, 7–9 June 2007, pp. 1620–1631
23. Cranefield, S., Pan, J.: Bridging the gap between the model-driven architecture and ontology engineering. ScienceDirect (2007)
24. http://neo4j.com/

Time Series Methods for Synthetic Video Traffic

Christos Katris and Sophia Daskalaki

Abstract The scope of this paper is the creation of synthetic video traffic using time series models. Firstly, we discuss the procedure for creating video traffic with FARIMA models. However, the created traffic displays the LRD characteristic of real traffic, but underestimates its moments (mean, sd, skewness and kurtosis). We present two approaches for improving the popular FARIMA model for the creation of synthetic traffic. The first approach is to apply FARIMA models with heavy-tailed errors for traffic creation. The second is a two step procedure, where we build a FARIMA model with normal innovations and then we provide a statistical transformation for its projection in order to catch a desired marginal probability distribution. Using this procedure we approximated Student t and LogNormal as marginal distributions. The above procedures are applied to the performance evaluation of three real VBR traces.

Keywords Synthetic traffic · FARIMA · Non normal marginals · Lognormal · Performance evaluation · VBR traffic

1 Introduction

Nowadays VBR traffic is a major portion of Internet traffic. Video streaming, high definition TV, and video conferencing are some of the widely used applications today and the traffic they create still requires attention. Designing models that generate traffic with their characteristics is valuable for many purposes such as forecasting, bandwidth allocation or performance evaluation. It is known that most forms of Internet traffic carry certain statistical characteristics, such as

C. Katris (✉) · S. Daskalaki
Department of Electrical & Computer Engineering, University of Patras, Patras, Greece
e-mail: chriskatris@upatras.gr

S. Daskalaki
e-mail: sdask@upatras.gr

© Springer International Publishing Switzerland 2016   99
T.B. Nguyen et al. (eds.), *Advanced Computational Methods
for Knowledge Engineering*, Advances in Intelligent Systems
and Computing 453, DOI 10.1007/978-3-319-38884-7_8

self-similarity and Long Range Dependence (LRD) [3, 16], quite often co-existing with Short Range Dependence (SRD) [19]. Stochastic models that capture these characteristics have been developed and applied quite successfully mainly to predict video traffic data. Amongst them FARIMA models [9, 10] are the most popular and have been extensively used for predicting video traffic [6, 13]. Moreover, one can find applications of FARIMA models for traffic performance evaluation in [4, 8].

Following [1], a successful traffic model can be evaluated by its ability to approximate statistical characteristics of real video trace such as probability density function (pdf), mean, variance, peak, coefficient of variation and autocorrelation. Moreover, the model should be able to produce synthetic traffic imitating a wide range of video sources, and all these with reasonable computational effort. Briefly, a successful traffic model not only has to capture the characteristics of video traffic, but also to predict accurately its network performance (e.g. queuing behavior of the network) [20].

We notice that a classical FARIMA model may capture basic characteristics such as mean and variance, correlation structure (i.e. short range dependence-SRD and long range dependence-LRD) but fails to capture the pdf of real traffic, thus to produce synthetic traffic similar to the real trace. There are papers such as [12, 14], where heavy-tailed innovations instead of Gaussian were used. Such approaches may sometimes lead to more accurate results with additional computational effort.

In this paper, we propose two procedures that may improve FARIMA model for generating synthetic traffic. The first approach is the FARIMA model with non normal (Student's t or Normal Inverse Gaussian) errors. Following this approach we succeed in generating synthetic traffic which displays statistics closer to real traffic, however the pdf of the real traffic cannot be captured accurately. The second is our proposed approach for creating synthetic traffic and uses a FARIMA model for generating traffic and also specify a pdf model which describes accurately the real traffic trace. Then we perform a projection of the synthetic traffic to the distribution of the original trace through percentiles. The goal is to maintain the dependence structure which was captured by the FARIMA model and at the same time capture the pdf of the real trace. We can approximate any marginal distribution (pdf regardless of time) we wish, but in this work we create traffic with Student t and LogNormal marginal distributions which need only 2 parameters for their characterization. The LogNormal distribution especially, is suitable for the description of internet traffic data, as it can describe data with heavy-tails and skewness. Synthetic trace from such a traffic model seems capable to capture queuing dynamics of real traffic. We notice that this approach is valid when the time series are stationary, which means that the marginal distribution is the same over time and the parameters are constant. In addition, according to [4] the pdf of traffic plays the most important role to queuing behavior, while SRD has a lower effect and LRD plays the least important role. According to this hierarchy, this second approach seems promising for accurate results.

For the remaining of the paper, Sect. 2 describes and analyzes the different procedures for creating synthetic traffic, Sect. 3 describes the performance evaluation analysis, while in Sect. 4 we test the procedures to real traffic traces, which are aggregated at the level of 12 frames, and compare their ability to capture queuing dynamics. Finally, in Sect. 5 the conclusions and main results of this work are highlighted.

2 Generation of Synthetic Video Traffic

Modeling Internet traffic is known to require self-similar processes due to the LRD dependence structures it displays, and occasionally autoregressive models due to co-existence of SRD. As a result FARIMA models have been widely used since they can capture both properties at the same time.

A FARIMA time series model is formulated as:

$$\Phi_p(L)\,(1-L)^d X_t = \Theta_q(L)\varepsilon_t, \text{ where } L \text{ is the lag operator,}$$
$$\Phi_p(L) = 1 - \varphi_1 L - \ldots - \varphi_p L^p \text{ and } \Theta_q(L) = 1 + \theta_1 L + \ldots + \theta_q L^q. \tag{1}$$

$$\text{Moreover, } (1-L)^d = \sum_{j=0}^{\infty} \binom{d}{j}(-L)^j,$$

and the error terms ε_t are typically assumed to be Normal white noise with mean 0 and variance σ^2.

FARIMA models have been used extensively for forecasting video traffic and other time series that exhibit LRD. Examples of papers which model data with the use of FARIMA model are [6, 15, 19]. When it comes to generating traffic for performance evaluation, however, apparently it cannot capture the marginal distribution and queuing dynamics of the real traffic. The fitting of a FARIMA model may follow the procedure that is described in [15]. Briefly, the series are demeaned first and then the fractional parameter d is estimated using the Geweke and Porter-Hudak estimator. We then decide the order (p, q) of the best ARMA model for the data using BIC (Bayesian Information Criterion). For computational complexity reasons, we restrict orders (p and q) to be less or equal than five. The selected order is the order for the FARIMA model too. Lastly, the remaining parameters (φ_i, θ_i) are estimated using a Maximum Likelihood procedure.

At this point we present two approaches for generating synthetic Internet traffic, which use as base simulated data from a FARIMA model and then through suitable transformations we attempt to add characteristics that are detected in real video traces.

2.1 FARIMA Models with Non Normal Errors

At first, we simulate observations from the fitted FARIMA model. Then, we ignore the data which are smaller than a lower threshold we define. Using this approach, we avoid negative and very low values which do not represent real traffic. After these two steps, we obtain synthetic traffic that approximates the mean and variance of the real trace, as well as the dependence structure i.e. SRD and LRD. The weakness in this approach lies to the fact that real traffic deviates from normality, something that the conventional FARIMA model assumes as a prerequisite. An alternative approach is to consider FARIMA models with non normal errors. For this work, the considered error distributions are the location-scale Student t and the Normal Inverse Gaussian.

In the first case, we assume that the errors follow Student-t distribution in its location-scale version:

$$f(x; a, \beta, v) = \frac{\Gamma\left(\frac{v+1}{2}\right)}{\sqrt{\beta v \pi} \Gamma\left(\frac{v}{2}\right)} \left[1 + \frac{(x-\alpha)^2}{\beta v}\right]^{-\left(\frac{v+1}{2}\right)}$$

with location parameter α, scale parameter β and shape parameter v, while Γ is the Gamma function. The location parameter α is the mean which is 0 and the variance is $\text{Var}(x) = \frac{\beta v}{(v-2)}$.

In the second case, the errors are assumed to follow a subclass of Generalized Hyperbolic distribution, the Normal Inverse Gaussian (NIG) distribution [2]:

$$f(x; \mu, \alpha, \beta, \delta, \gamma) = \frac{\alpha \delta K_1\left(a\sqrt{\delta^2 + (x-\mu)^2}\right)}{\pi\sqrt{\delta^2 + (x-\mu)^2}} e^{\delta\gamma + \beta(x-\mu)}$$

with location parameter μ, tail heaviness parameter α, asymmetry parameter β, scale parameter δ and $\gamma = \sqrt{a^2 - \beta^2}$. Also, K_1 is a modified Bessel function of the third kind.

Using the above models as error distributions, we obtain synthetic traffic that approximates not only the dependence structure i.e. SRD and LRD, but also the heavy tails of the real traffic.

2.2 Approximating the Marginal Distribution of the Data

From studying the marginal distributions of Internet traffic traces it is certain that they display heavy tails and high variability. A probability distribution, which carries such characteristics, is therefore essential for any procedure that attempts to generate synthetic traffic. In addition, a model with small number of parameters, such as the LogNormal(LN) distribution suggested in [5] is a preferable choice. LogNormal is the distribution of a continuous random variable whose logarithm is normally distributed. It requires two parameters, μ and σ^2, the mean and variance of its logarithm. The LogNormal distribution display right skewness and heavy tails and its pdf is:

$$f(x) = \frac{1}{x\sigma\sqrt{2\pi}} e^{-\frac{(\ln(x)-\mu)^2}{2\sigma^2}} \tag{2}$$

More details about the LogNormal distribution can be found in [18]. Except from the LogNormal distribution, we project the FARIMA model to the Student t distribution which was described above.

We now present a 2 stage procedure which leads to the generation of synthetic Internet traffic by projecting simulated FARIMA data to a LogNormal distribution. Such generation appears to approximate reasonably well the mean, variance and also shape of the marginal, as well as the dependence structure of the real traffic trace.

Like previously, at the first stage N traffic observations $(y_i, i = 1, \ldots, N)$ are simulated from the FARIMA model. Based on the assumptions of this model the simulated data will carry SRD and LRD dependencies, however, their marginal distribution will be approximately normal $(y_i \sim N(\mu, \sigma))$, with mean μ and standard deviation σ that are estimated from the real traffic trace. On the contrary, we assume that the original data follow a LogNormal distribution with parameters the mean and variance of the logarithms of the original data. Therefore, at the 2nd stage we consider a transformation of the simulated data using percentiles, which allows to project the created traffic of the FARIMA model, to the considered LogNormal distribution.

More specifically, the probabilities p_i's that correspond to the y_i's are all calculated using the Normal distribution (i.e. $p_i = F_N(y_i)$). Then the percentiles x_i that correspond to the p_i's on the horizontal axis of the LogNormal distribution are found, using the inverse cumulative function: $x_i = F_{LN}^{-1}(p_i)$. The new values x_i are distributed according to $LN(\hat{\mu}_{\log y_i}, \hat{\sigma}_{\log y_i})$ and constitute the synthetic traffic that as it turns out emulate quite nicely the behavior of the real traces, at least in a queuing situation.

The motivation for this transformation is that when we simulate a sequence of data from a model with certain dependence structure properties, the simulated data will exhibit the same structure. However, the simulated data are certain points on the horizontal axis of their CDF that correspond to their marginal distribution (for example Normal). A sequence of points on the horizontal axis of the desired CDF distribution that corresponds to the same probabilities as the simulated ones will exhibit approximately the dependence structure of the original sequence. So, with the use of such transformation, we can obtain points with a desired marginal distribution and also dependence structure similar to that of the simulated data.

The procedure just presented assumes that for start there is a real trace where a FARIMA model is fit. In case of a new video however, this will not be possible. So for the task of evaluation of a new video, we can assume a FARIMA(1, 1) as reference model, with coefficients $p_1 = 0.5, q_1 = 0.25$ and $d = 0.3$, which corresponds to H = 0.8, quite a realistic assumption for real traces. With such choice the moving average component of the model will carry a less important role than its autoregressive counterpart. Still we will have to specify a mean and a variance of the new trace. Then, we can apply the proposed transformation, in order to capture the LogNormal marginal of the traffic. The parameters of the corresponding LogNormal distribution are estimated through the MCMC method. In the experimental analysis presented in Sect. 4 this procedure will be referred as "Projected FARIMA (1, 1) to LN".

3   Performance Evaluation Analysis

The procedures presented in Sect. 2 focus on generating traffic that carry characteristics of real video traffic and play significant roles in performance evaluation. We now describe a possible queuing framework where we can evaluate the performance of traffic traces.

Let's consider a single server FIFO queue with deterministic service rate and finite buffer. The queue is fed with traffic which is either from the real traces, or traffic generated from one of the procedures described previously. At each time period traffic arrives carrying a certain load (in KB) and requests the corresponding bandwidth. If the traffic load exceeds the available service capacity, the excess is placed in the buffer; however if the buffer size is also exceeded then loss is realized. The traffic that should be served at time t (TR_t) is the new traffic that arrives at time t (NT_t) plus the buffered traffic up until time $t - 1$ (QT_{t-1}).

$$TR_t = NT_t + QT_t \tag{3}$$

The performance measure that is used for the queuing system under study is the loss probability $\Pr[TR_t > SC + BS]$, where SC is the service capacity and BS the buffer size.

The goal for each traffic model is to be as close as possible to the real traffic trace, with the closeness referring not only to the descriptive statistics, but also to the dependence structure characteristics, i.e. SRD and LRD, and also its queuing behavior for a wide range of service rates and buffer sizes.

4   Modeling of Real Traces

We now apply the procedures described in Sect. 2 to three VBR traces in order to generate synthetic traffic that can mimic the corresponding real traces in queuing situations. The traces used are publically available from TU-Berlin [7] and have been used extensively for different studies. These are H.263 traces with VBR target rate. For uniformity purposes, each trace was aggregated so to give KB per 12 frames and from each trace we used a dataset of 4000 12-frame samples.

Table 1 displays the descriptive statistics for each dataset, i.e. number of data considered, mean, standard deviation, skewness, kurtosis and coefficient of

Table 1 Descriptive statistics of data

Trace	Count	Mean	Sd	Skewness	Kurtosis	CV	Stationarity* ADF test (p-value)
Silence of the Lambs	4000	32.108	31.343	3.256	19.438	0.976	−8.9462 (<0.01)
Mr Bean	4000	31.608	16.977	1.922	10.453	0.537	−8.8663 (<0.01)
Star Trek	4000	17.640	12.799	2.518	14.158	0.725	−9.9472 (<0.01)

*stationarity is the alternative hypothesis

Table 2 Characteristics of the dependence structure

Trace	Autocorrelation Ljung-Box (p-value)	Hurst exponent (H)
Silence of the Lambs	3496.505 (<0.01)	0.9173369
Mr Bean	2779.975 (<0.01)	0.7148368
Star Trek	3026.608 (<0.01)	0.8155737

variation. These statistics are static and ignore the effect of time, something which is acceptable because data can be assumed stationary according to ADF test. All three traces,are shown to be right skewed and leptokurtic. Comparing the coefficients of variation (CV), one may observe that variability is different among traces, with that of *Mr. Bean* carrying the lowest and that of the *Silence of the Lambs* carrying the highest. The CV can be seen as a measure of difficulty in approximating successfully a trace. Larger CV implies larger variability and thus additional difficulty to approximate the queuing behavior of the traffic trace. Table 2 presents characteristics that describe the dependence structure in the data. More specifically, autocorrelation is checked using the Ljung-Box test and long memory is measured using the Hurst exponent (H), calculated via the R/S method [11, 17]. Since $0.75 < H < 1.0$ and the p-values of Ljung-Box tests were all very small, we conclude that all three traces exhibit strong long-memory and significant autocorrelation. Next FARIMA models were fit to all traces and Table 3 gives the final values for the parameters of the fitted models. Specifically, the order p for the autoregressive, q for the moving average components, and d for the corresponding fractional parameter are shown.

Table 3 Parameters for the fitted FARIMA models

Trace	Order (p, q)	Fractional parameter (d)
Silence of the Lambs	(3, 1)	0.4711255
Mr Bean	(1, 4)	0.3289572
Star Trek	(1, 3)	0.3131650

Given the FARIMA models created from the fitting procedure we now create synthetic traffic with the procedures which were described in Sect. 2 (i.e. FARIMA with Student t and Normal Inverse Gaussian errors and the 2-stage approximation of the marginal distribution of the data). For each model, traffic traces with 100,000 points were created initially, but we keep only the values which exceed the minimum value of each trace. Table 4 presents the number of data of the created traffic, the statistical characteristics, i.e. mean, standard deviation, skewness, kurtosis and Hurst exponent for LRD, for the real data and for the data generated from the traffic models. We can see that all three synthetic traces indicate existence of long memory, although the estimated Hurst exponents are slightly different. We notice that all created traces display significant autocorrelation according to the Ljung-Box statistical test.

According to the results, the estimated locations (means) and dispersions (standard deviations) are relatively close, with the FARIMA with student t errors and especially the 2-stage FARIMA to LN distribution to be closer than the other models. The classical FARIMA model underestimates these parameters mainly for the *Silence of the Lambs*, less for *Mr. Bean* and is close for *Star Wars*. A major discrepancy of the classical FARIMA can also be detected in the shape parameters (i.e. skewness and kurtosis) for all traces. Such an observation indicates that the

Table 4 Comparison of real with generated traces

Trace	# of data	Mean	Sd	Skewness	Kurtosis	H
Silence of the Lambs	4000	32.108	31.343	3.256	19.438	0.9173
FARIMA	16845	22.820	17.605	1.158	4.242	0.8790
FARIMA—Student t	25055	33.801	33.565	3.381	26.634	0.9403
FARIMA—NIG	17774	32.790	30.473	2.030	9.451	0.7380
2 stage FARIMA— Student t	86990	39.611	25.012	1.258	6.316	0.8536
2 stage FARIMA—LN	99997	31.750	29.310	3.010	18.676	0.8424
Mr Bean	4000	31.608	16.977	1.922	10.453	0.7148
FARIMA	91527	27.518	14.572	0.451	2.865	0.6937
FARIMA—Student t	86672	28.983	17.544	1.166	7.261	0.9094
FARIMA—NIG	95283	28.101	14.152	0.795	4.614	0.9124
2 stage FARIMA— Student t	96452	33.164	14.985	0.970	7.962	0.7601
2 stage FARIMA—LN	100000	31.813	17.879	1.896	10.129	0.7938
Star Trek	4000	17.640	12.799	2.518	14.158	0.8156
FARIMA	79343	17.002	10.172	0.619	3.063	0.7022
FARIMA—Student t	75075	19.128	14.356	2.861	31.820	0.9230
FARIMA—NIG	85898	17.423	10.733	0.953	4.603	0.7516
2 stage FARIMA— Student t	92775	19.636	10.726	1.266	9.564	0.6448
2 stage FARIMA—LN	100000	17.876	13.889	2.905	21.605	0.7822

marginal distribution of the traffic generated with this procedure differs significantly from the actual. The FARIMA with NIG errors also underestimates skewness and kurtosis, while FARIMA with student t errors cannot capture accurately these characteristics. The 2-stage FARIMA student t also underestimates 3rd and 4th moments (skewness and kurtosis). It is interesting to notice that the 2-stage FARIMA to LN approximates the descriptive characteristics of the real traffic better than the other models. For the dependence structure, all models display significant autocorrelation. The classical *FARIMA* model is closest to the actual long memory as displays the Hurst exponent, but traffic from all models displays long-memory.

The next step is to measure the queuing performance of the traces and monitor the ability of the proposed traffic modeling procedure to create traffic that can capture the queuing dynamics of real VBR traffic traces. For each real trace we consider a FIFO queuing system with service capacity (SC) equal to the 80 % of the sample mean plus one standard deviation $(\bar{x} + s)$. Then we estimate the queuing performance for different buffer sizes, from 50 up to 450 KBs. Furthermore, we consider a fixed buffer size equal to 100 KB and estimate the queuing performance for different service rates. For each trace we consider initially value of $(\bar{x} + s)$ as service capacity and subsequently considered several fractions of that value.

Figure 1 displays the queuing performance of the real trace, indicated as "Loss-Trace", and those of the models which used for simulating synthetic traffic, FARIMA models with Student t and NIG errors and 2 step FARIMA models with projection to Student t and LogNormal marginal distributions. We observed that there is an increasing difficulty in capturing the queuing behavior of the real traces, as CV increases. For that reason, we have 3 traces with different levels of difficulty in capturing queuing dynamics. The Silence of the Lambs trace is the most difficult to approximate its queuing behavior and has the highest CV, while Mr. Bean is approximated better than the rest and the trace carried the lowest CV. Star Trek trace represents an intermediate situation.

In Fig. 1 we observe that the loss probability achieved using the traffic generated by alternative models is more accurate than that of the classic FARIMA model, which fails to accurately capture the queuing behavior of real traffic across buffer sizes in all cases, while for service rates was accurate only for *Mr. Bean* trace. All other models displayed a better overall approach, however only the 2-stage FAR-IMA to LN model was always close to the real trace queuing behavior for all cases. Even with the most difficult situation for the *Silence of the Lambs* trace, the model estimated quite accurately the queuing dynamics of the real traffic. As for the other models, the 2-stage FARIMA-Student t failed to describe traffic of the *Silence of the Lambs* trace, FARIMA with NIG errors failed to capture the behavior of actual traffic for the *Mr. Bean* trace and FARIMA with Student t errors failed to describe accurately the behavior of *Star Trek* trace. Overall, the approaches improved the ability of classical FARIMA model and the best approach is certainly the 2-stage FARIMA to LN.

The 2-stage FARIMA to LN model found to be accurate for measuring the performance of an existing video. In case of a new video, we only need to know or to estimate the mean value and the standard of the video trace. Then, we apply a

Silence of The Lambs

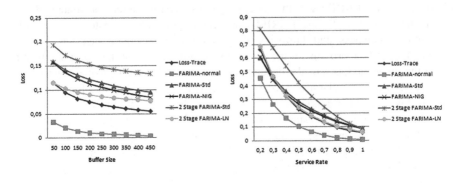

Mr. Bean

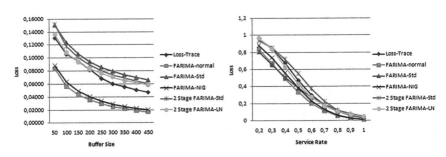

Star trek

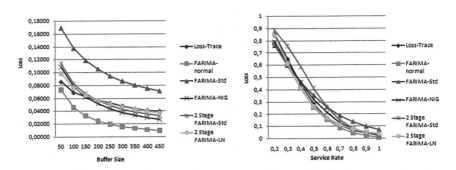

Fig. 1 Performance evaluation of models

'Projected FARIMA (1, 1) to LN' model, as described in Sect. 2.2. Figure 2 displays the performance of the model for capturing real traffic compared with the classical *FARIMA* model and the *2-stage FARIMA to LN* model.

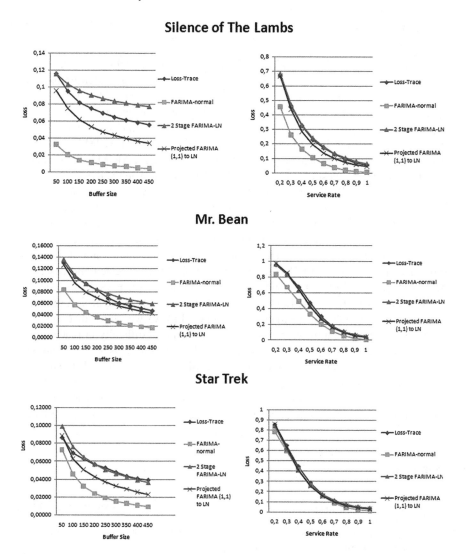

Fig. 2 Performance evaluation for a new video

From Fig. 2, we observe that the '*Projected FARIMA (1, 1) to LN*' model is a very accurate approximator for VBR traffic traces, only a little bit poorer than 2 stage FARIMA to LN, as we expected, because the model does not use fitting, but displays clearly better results than classical FARIMA model for *Mr. Bean* and *Star Trek* traces. This is evidence that the 2-stage FARIMA to LN model can be used successfully for the evaluation of new videos.

5 Conclusions

In this paper we studied the problem of generating synthetic traffic that behaves like a real video trace in a queuing situation. Such a generator could then be used for performance analysis evaluation and network design. Since video traces carry specific characteristics that have to do either with the marginal distribution or the dependence structure and autocorrelation function of the time series three procedures are examined, all based on FARIMA models (classical FARIMA, FARIMA with non normal errors and a procedure which projects FARIMA to a desired distribution). We suggest that a traffic model may approximate the dynamics of real traffic if it is capable of approximating its queuing behavior for a wide range of buffer sizes and service rates. Using three real VBR traces, aggregated into 12-frame batches we evaluated and compared the performance of the three procedures. The study indicates that a traffic generating procedure can be successful if it matches location, dispersion and shape statistics of the traffic, and at the same time maintains SRD and LRD. Such a procedure has been detected and described here (2-stage FARIMA to LN) and involves the projection of FARIMA sequences to a desired distribution—the LogNormal used as an appropriate approach for video traffic—using inverse transformation technique. Lastly, we show that the CVs calculated from the real traces can be used as measure of difficulty in matching its queuing behavior. The proposed procedure followed quite closely the examined traces and found to perform better than the classical FARIMA model. Finally, for the challenging task of evaluating the performance of a new video trace, we found from empirical results that a FARIMA (1, d, 1) can be used as reference and then apply the proposed transformation using only an estimate of the mean and variance.

References

1. Alheraish, A.A.: Autoregressive video conference models. Int. J. Netw. Manage. 14(5), 329–337 (2004)
2. Barndorff-Nielsen, O.: Hyperbolic distributions and distributions on hyperbolae. Scandinavian J. Stat.151–157 (1978)
3. Beran, J., Sherman, R., Taqqu, M.S. Willinger, W.: Variable bit-rate video traffic and long range dependence. IEEE Trans. Commun. 43(2/3/4), 1566–1579 (1995)
4. Casilari, E., Reyes, A., Diaz-Estrella, A., Sandoval, F.: Characterization and modelling of VBR video traffic. Electron. Lett. 34(10), 968–969 (1998)
5. Casilari, E., Reyes, A., Diaz-Estrella, A., Sandoval, F.: Classification and comparison of modelling strategies for vbr video traffic. In: Proceedings of International Teletraffic Congress (ITC-16) June 1999
6. Corradi, M., Garroppo, R.G., Giordano, S., Pagano, M..: Analysis of f-ARIMA processes in the modeling of broadband traffic. In: ICC'01, vol. 3, pp. 964–968 (2001)
7. Fitzek, F.H.P., Reisslein, M.: MPEG-4 and H.263 video traces for network performance evaluation. IEEE Netw. Mag. 15(6), 40–54 (2001)

8. Garroppo, R.G., Giordano, S.: Comparison of LRD and SRD traffic models for the performance evaluation of finite buffer systems. IEEE Int. Conf. Commun. ICC **2001**(9), 2681–2686 (2001)
9. Granger, C.W.J., Joyeux, R.: An introduction to long-memory time series models and fractional differencing. J. Time Ser. Anal. **1**, 15–30 (1980)
10. Hosking, J.R.M.: Fractional differencing. Biometrika **68**, 165–176 (1981)
11. Hurst, H.E.: Long-term storage capacity of reservoirs. Trans. Am. Soc. Civ. Eng. **116**, 770–808 (1951)
12. Ilow, J.: Forecasting network traffic using FARIMA models with heavy tailed innovations. In: Proceedings of IEEE International Conference on Acoustics, Speech, and Signal Processing (ICASSP'00), pp. 3814–3817 (2000)
13. Jin, Z., Shu, Y., Liu, J., Oliver, W.: Prediction-based bandwidth allocation for VBR traffic. Trans. Tianjin Univ. **7**(4), 221–225 (2001)
14. Katris, C., Daskalaki, S.: Combining Time Series Forecasting Methods for Internet Traffic. In: Stochastic Models, Statistics and Their Applications, pp. 309–317. Springer International Publishing (2015)
15. Katris, C., Daskalaki, S.: Comparing forecasting approaches for Internet traffic. Expert Syst. Appl. **42**(21), 8172–8183 (2015)
16. Leland, W.E., Taqqu, M.S., Willinger, W., Wilson, D.: On the self-similar nature of Ethernet traffic (extended version). IEEE/ACM Trans. Netw. **2**, 1–15 (1994)
17. Mandelbrot, B.: Statistical methodology for non-periodic cycles: from the covariance to R/S analysis. Ann. Econ. Soc. Measur. **1**, 257–288 (1972)
18. NIST/SEMATECH: Engineering Statistics Handbook—LogNormal Distribution (2006)
19. Shu, Y., Jin, Z., Zhang, L., Wang, L., Oliver, W., Yang, W.: Traffic prediction using FARIMA models. IEEE Int. Conf. Commun. **2**, 891–895 (1999)
20. Tanwir, S., Perros, H.: A survey of VBR video traffic models. IEEE Commun. Surv. Tut. **15**(4), 1778–1802 (2013)

Part III
Topics on Discrete Mathematics

A Constraint Solver for Equations over Sequences and Contexts

Mariam Beriashvili and Besik Dundua

Abstract In this paper we propose a solving algorithm for equational constraints over unranked terms, contexts, and sequences. Unranked terms are constructed over function symbols which do not have fixed arity. For some function symbols, the order of the arguments matters (ordered symbols). For some others, this order is irrelevant (unordered symbols). Contexts are unranked terms with a single occurrence of hole. Sequences consist of unranked terms and contexts. Term variables stand for single unranked terms, sequence variables for sequences, context variables for contexts, and function variables for function symbols. We design an terminated and incomplete constraint solving algorithm, and indicate a fragment for which the algorithm is complete.

1 Introduction

Unranked terms are built over function symbols which do not have a fixed arity (unranked symbols). They are nearly ubiquitous in XML-related applications [18]. They model variadic procedures used in programming languages [2, 22, 23]. Moreover, they appear in rewriting [10], knowledge representation [9, 20], theorem proving [12, 14], program synthesis [3], just to name a few.

When working with unranked terms, it is a pragmatic necessity to consider variables which can be instantiated by a finite sequences of terms (called sequences). Such variables are referred to as sequence variables. An example of an unranked term is $f(\overline{x}, f, x, \overline{y})$, where f is an unranked function symbol, $\overline{x}$ and $\overline{y}$ are sequence variables, and x is a usual term variable which can be instantiated by a single term. We can match this term, e.g., to the term $f(f, a, f, b)$ in two different ways, with the substitutions $\{\overline{x} \mapsto (\,), x \mapsto a, \overline{y} \mapsto (f, b)\}$ and $\{\overline{x} \mapsto (f, a), x \mapsto f, \overline{y} \mapsto b\}$, where $(\,)$ is the empty sequence and (f, a) is a sequence consisting of two terms f and a. Terms are singleton sequences.

M. Beriashvili · B. Dundua (✉)
Vekua Institute of Applied Mathematics, Tbilisi State University, 0183 Tbilisi, Georgia
e-mail: bdundua@gmail.com

© Springer International Publishing Switzerland 2016
T.B. Nguyen et al. (eds.), *Advanced Computational Methods for Knowledge Engineering*, Advances in Intelligent Systems and Computing 453, DOI 10.1007/978-3-319-38884-7_9

Sequences can be concatenated to each other. In this way, sequences can "grow horizontally" and sequence variables help explore it by filling gaps between siblings. However, such a concatenation has limited power, since it does not affect the depth of sequences, i.e., it does not permit sequences "to grow vertically". To address this problem, Bojańczyk and Walukiewicz [1] introduced forest algebras, where alongside sequences (thereby called forests), context also appears. Contexts are sequences with a single occurrence of the hole symbol placed in some leaf. Contexts can be composed by putting one of them in the hole of the other. Moreover, context can apply to a sequence by putting it into the hole, resulting in a sequence. One can introduce context variables to stand for such contexts, and function variables to stand for function symbols.

Reasoning about sequences gives rise to constraints which should be solved. This turned out to be quite a difficult task. Even if we consider unification problems, in the presence of sequence variables or context variables alone they are infinitary [13, 17]. They both generalize word unification [19]. Several finitary fragments and variants of context and sequence unification problems have been identified. Solving in a theory which combines both context and sequence variables is relatively less studied, with the exception of context sequence matching [15] and its application in rule-based programming [8].

We may have function symbols whose argument order does not matter (unordered symbols): A kind of generalization of the commutativity property to unranked terms. The programming language of Mathematica [23] is an example of successful application in programming of both syntactic and equational unranked pattern matching (including unordered matching) algorithms with sequence variables.

Various forms of constraint solving are in the center of declarative programming paradigms. Unification is the main computational mechanism for logic programming. Matching plays the same role in rule-based and functional programming. Constraints over special domains are in the heart of constraint logic programming languages.

In [7] we have studied a constraint solver for unranked sequences built over ordered and unordered function symbols. In this paper, we generalize this approach by combining contexts and unranked sequences in a single framework. Such a language is rich, possesses powerful means to traverse trees both horizontally and vertically in a single or multiple steps, and allows the user to naturally express data structures (e.g., trees, sequences, multisets) and to write code concisely. We propose a solving algorithm for constraints over terms, contexts, and sequences. The algorithm works on the input in disjunctive normal form and transforms it to the partially solved form. It is sound and terminating. The latter property naturally implies that the solver is incomplete for arbitrary constraints, because the problem it solves is infinitary: There might be infinitely many incomparable solutions to constraints that involve sequence and context variables, see, e.g., [11, 17]. However, there are fragments of constraints for which the solver is complete, i.e., it computes all the solutions. One of such fragments is the so called the well-moded fragment [7], where variables on one side of equations (or in the left hand side of the membership atom) are guaranteed to be instantiated with ground expressions at some point. This effectively reduces constraint solving to sequence matching and context matching (which are known to be NP-complete [16, 21]), plus some early failure detection rules.

2 The Language

The alphabet $\mathcal{A}$ contains the sets of *term variables* $\mathcal{V}_T$, *sequence variables* $\mathcal{V}_S$, *function variables* $\mathcal{V}_F$, *context variables* $\mathcal{V}_C$, *unranked unordered function symbols* $\mathcal{F}_u$ and *ordered function symbols* $\mathcal{F}_o$. All these sets are assumed to be mutually disjoint. Henceforth, we shall assume that the symbols: x, y and z range over $\mathcal{V}_T$; $\bar{x}, \bar{y}, \bar{z}$ over $\mathcal{V}_S$; X, Y, Z over $\mathcal{V}_F$; $X_\bullet, Y_\bullet, Z_\bullet$ over $\mathcal{V}_C$; f_u, g_u, h_u over $\mathcal{F}_u$ and f_o, g_o, h_o over $\mathcal{F}_o$. Moreover, *function symbols* denoted by f, g, h are elements of the set $\mathcal{F} = \mathcal{F}_u \cup \mathcal{F}_o$, a *variable* is an element of the set $\mathcal{V} = \mathcal{V}_T \cup \mathcal{V}_S \cup \mathcal{V}_F \cup \mathcal{V}_C$ and a *functor* F is a common name for a function symbol or a function variable. The alphabet also contains the *special constant* $\bullet$, the *propositional constants* true, false, the *logical connectives* $\neg, \wedge, \vee, \Rightarrow, \Leftrightarrow$, the *quantifiers* $\exists, \forall$ and the *binary equality predicate* $\doteq$.

Definition 1 We define inductively *terms, sequences, contexts* and other syntactic categories over $\mathcal{A}$ as follows:

$$
\begin{array}{llll}
t & ::= x \mid F(S) \mid X_\bullet(t) & & \text{Term} \\
T & ::= t_1, \ldots, t_n & (n \geq 0) & \text{Term sequence} \\
\tilde{s} & ::= t \mid \bar{x} & & \text{Sequence element} \\
S & ::= \tilde{s}_1, \ldots, \tilde{s}_n & (n \geq 0) & \text{Sequence} \\
C & ::= \bullet \mid F(S, C, S) \mid X_\bullet(C) & & \text{contexts}
\end{array}
$$

For readability, we put parentheses around sequences, writing, e.g., $(f(a), \bar{x}, b)$ instead of $f(a), \bar{x}, b$. The empty sequence is written as $(\ )$. Besides the letter t, we use also r and s to denote terms. Two sequences are *disjoint* if they do not share a common element. For instance, $(f(a), x, b)$ and $(f(x), f(b, f(a)))$ are disjoint, whereas $(f(a), x, b)$ and $(f(b), f(a))$ are not.

A context C may be applied to a term t (resp. context C'), written $C[t]$ (resp. a context $C[C']$), and the result is the term (resp. context) obtained from C by replacing the hole $\bullet$ with t (resp. with C'). Besides the letter C, we use also D to denote contexts.

The set of terms is denoted by $\mathcal{T}(\mathcal{F}, \mathcal{V})$ and the set of contexts is denoted by $\mathcal{C}(\mathcal{F}, \mathcal{V})$.

Definition 2 A *formula* over the alphabet $\mathcal{A}$ is defined inductively as follows:

1. true and false are formulas.
2. If t and r are terms, then $t \doteq r$ is a formula.
3. If C and D are contexts, then $C \doteq D$ is a formula.

4. If $\mathbf{F}_1$ and $\mathbf{F}_2$ are formulas, then so are $(\neg\mathbf{F}_1)$, $(\mathbf{F}_1 \vee \mathbf{F}_2)$, $(\mathbf{F}_1 \wedge \mathbf{F}_2)$, $(\mathbf{F}_1 \Rightarrow \mathbf{F}_2)$, and $(\mathbf{F}_1 \Leftrightarrow \mathbf{F}_2)$.
5. If $\mathbf{F}$ is a formula and $v \in \mathcal{V}_S$, then $\exists v.\mathbf{F}$ and $\forall v.\mathbf{F}$ are formulas.

The formulas defined by the items (2) and (3) are called *primitive constraints*. A *constraint* C is an arbitrary formula built over true, false and primitive constraints.

A *substitution* is a mapping from term variables to terms, from sequence variables to sequences, from function variables to functors, and from context variables to contexts, such that all but finitely many term, sequence, and function variables are mapped to themselves, and all but finitely many context variables are mapped to themselves applied to the hole. Substitutions extend to terms, sequences, contexts, formulas. The *sets of free and bound variables* of a formula $\mathbf{F}$, denoted fvar($\mathbf{F}$) and bvar($\mathbf{F}$) respectively, are defined in the standard way as can be seen in [6].

3 Semantics

For a given set S, we denote by S^* the set of finite, possibly empty, sequences of elements of S, and by S^n the set of sequences of length n of elements of S. Given a sequence $\bar{s} = (s_1, s_2, \dots, s_n) \in S^n$, we denote by $perm(s)$ the set of sequences $\{(s_{\pi(1)}, s_{\pi(2)}, \dots, s_{\pi(n)}) \mid \pi \text{ is a permutation of } \{1, 2, \dots, n\}\}$. The set of functions from a set S_1 to a set S_2 is denoted by $S_1 \longrightarrow S_2$. The notion $f : S_1 \longrightarrow S_2$ means that f belongs to $S_1 \longrightarrow S_2$.

A *structure* $\mathfrak{S}$ for a language $\mathcal{L}(\mathcal{A})$ is a tuple $\langle D, \mathcal{I} \rangle$ made of a non-empty *carrier set* of individuals D and an interpretation function $\mathcal{I}$ that maps each function symbol $f \in \mathcal{F}$ to a function $\mathcal{I}(f) : D^* \longrightarrow D$. Moreover, if $f \in \mathcal{F}_u$ then $\mathcal{I}(f)(s) = \mathcal{I}(f)(s')$ for all $s \in D^*$ and $s' \in perm(s)$. Given such a structure, we also define the operation $\mathcal{I}_C : (D^* \longrightarrow D) \longrightarrow D^* \longrightarrow D^* \longrightarrow (D \longrightarrow D) \longrightarrow (D \longrightarrow D)$ by $\mathcal{I}_C(\psi)(\bar{s}_1)(\bar{s}_2)(\phi)(d) := \psi(\bar{s}_1, \phi(d), \bar{s}_2)$ for all $\psi : D^* \longrightarrow D$, $\bar{s}_1, \bar{s}_2 \in D^*$, $d \in D$, and $\phi : D \longrightarrow D$.

A *variable assignment* for such a structure is a function with the domain $\mathcal{V}$ that maps term variables to elements of D; sequence variable to elements of D^*; function variables to functions in $D^* \longrightarrow D$; and context variables to functions in $D \longrightarrow D$.

The interpretations of our syntactic categories with respect to a structure $\mathfrak{S} = \langle D, \mathcal{I} \rangle$ and variable assignment ρ is shown below. The *interpretation* of simple sequences $[\![S]\!]_{\mathfrak{S},\rho}$ and of contexts $[\![C]\!]_{\mathfrak{S},\rho}$ are defined as follows:

$$[\![x]\!]_{\mathfrak{S},\rho} := \rho(x).$$

$$[\![f(S)]\!]_{\mathfrak{S},\rho} := \mathcal{I}(f)([\![S]\!]_{\mathfrak{S},\rho}).$$

$$[\![X(S)]\!]_{\mathfrak{S},\rho} := \rho(X)([\![S]\!]_{\mathfrak{S},\rho}).$$

$$[\![X_\bullet(t)]\!]_{\mathfrak{S},\rho} := \rho(X_\bullet)([\![t]\!]_{\mathfrak{S},\rho}).$$

$$[\![\overline{x}]\!]_{\mathfrak{S},\rho} := \rho(\overline{x}).$$

$$[\![(\tilde{s}_1, \ldots, \tilde{s}_n)]\!]_{\mathfrak{S},\rho} := ([\![\tilde{s}_1]\!]_{\mathfrak{S},\rho}, \ldots, [\![\tilde{s}_n]\!]_{\mathfrak{S},\rho}).$$

$$[\![\bullet]\!]_{\mathfrak{S},\rho} := Id_D.$$

$$[\![f(S_1, C, S_2)]\!]_{\mathfrak{S},\rho} := I_C(I(f))([\![S_1]\!]_{\mathfrak{S},\rho})([\![S_2]\!]_{\mathfrak{S},\rho})([\![C]\!]_{\mathfrak{S},\rho}).$$

$$[\![X(S_1, C, S_2)]\!]_{\mathfrak{S},\rho} := I_C(\rho(X))([\![S_1]\!]_{\mathfrak{S},\rho})([\![S_2]\!]_{\mathfrak{S},\rho})([\![C]\!]_{\mathfrak{S},\rho}).$$

$$[\![X_\bullet(C)]\!]_{\mathfrak{S},\rho} := \rho(X_\bullet) \odot [\![C]\!]_{\mathfrak{S},\rho}, \text{ where } \odot \text{ stands for composition.}$$

Note that terms are interpreted as elements of D, sequences as elements of D^*, and contexts as elements of $D \longrightarrow D$. We may omit ρ and write simply $[\![E]\!]_{\mathfrak{S}}$ for the interpretation of a variable-free (i.e., *ground*) expression E.

Formulas with respect to a structure $\mathfrak{S}$ and a variable assignment ρ are *interpreted* as follows:

$$\mathfrak{S} \models_\rho \text{ true.}$$

$$\text{Not } \mathfrak{S} \models_\rho \text{ false.}$$

$$\mathfrak{S} \models_\rho t_1 \doteq t_2 \text{ iff } [\![t_1]\!]_{\mathfrak{S},\rho} = [\![t_2]\!]_{\mathfrak{S},\rho}.$$

$$\mathfrak{S} \models_\rho C_1 \doteq C_2 \text{ iff } [\![C_1]\!]_{\mathfrak{S},\rho} = [\![C_2]\!]_{\mathfrak{S},\rho}.$$

Interpretation of an arbitrary formula with respect to a structure and a variable assignment is defined in the standard way. Also, the notions $\mathfrak{S} \models F$ for validity of an arbitrary formula F in $\mathfrak{S}$, and $\models F$ for validity of F in any structure are defined as usual.

An *intended structure* is a structure $\mathfrak{I}$ with a carrier set $\mathcal{T}(\mathcal{F})$ (the set of ground simple terms) and interpretation I defined for every $f \in \mathcal{F}$ by $I(f)(S) := f(S)$. It follows that $I_C(I(f))(S_1)(S_2)(C) := f(S_1, C, S_2)$. Thus, intended structures identify terms, sequences and contexts with themselves. Also, $[\![R]\!]_{\mathfrak{I}}$ is the same in all intended structures, and will be denoted by $[\![R]\!]$. Other remarkable properties of intended structures $\mathfrak{I}$ are: $\mathfrak{I} \models_\rho t_1 \doteq t_2$ iff $t_1\rho = t_2\rho$ and $\mathfrak{I} \models_\rho C_1 \doteq C_2$ iff $C_1\rho = C_2\rho$.

A ground substitution ρ is a *solution* of a constraint C if $\mathfrak{I} \models C\rho$ for all intended structures $\mathfrak{I}$.

4 Solved and Partially Solved Constraints

We say a variable is *solved* in a conjunction of primitive constraints $\mathcal{K} = c_1 \wedge \cdots \wedge c_n$, if there is a c_i, $1 \leq i \leq n$, such that

- the variable is x, c_i is $x \doteq t$, and x occurs neither in t nor elsewhere in $\mathcal{K}$, or
- the variable is $\bar{x}$, c_i is $\bar{x} \doteq S$, and $\bar{x}$ occurs neither in $\tilde{s}$ nor elsewhere in $\mathcal{K}$, or
- the variable is X, c_i is $X \doteq F$ and X occurs neither in F nor elsewhere in $\mathcal{K}$, or
- the variable is $X_\bullet$, c_i is $X_\bullet \doteq C$, and $X_\bullet$ occurs neither in C nor elsewhere in $\mathcal{K}$, or

In this case we also say that $\mathbf{c}_i$ is *solved in* $\mathcal{K}$. Moreover, $\mathcal{K}$ is called *solved* if for any $1 \leq i \leq n$, $\mathbf{c}_i$ is solved in it. $\mathcal{K}$ is *partially solved*, if for any $1 \leq i \leq n$, $\mathbf{c}_i$ is solved in $\mathcal{K}$, or has one of the following forms:

- $(\bar{x}, S_1) \doteq (\bar{y}, S_2)$ where $\bar{x} \neq \bar{y}$, $S_1 \neq (\,)$ and $S_2 \neq (\,)$.
- $(\bar{x}, S_1) \doteq (S, \bar{y}, S_2)$, where S is a sequence of terms, $\bar{x} \notin var(S)$, $S_1 \neq (\,)$, and $S \neq (\,)$. The variables $\bar{x}$ and $\bar{y}$ are not necessarily distinct.
- $f_u(S_1, \bar{x}, S_2) \doteq f_u(S_3, \bar{y}, S_4)$ where $(S_1, \bar{x}, S_2)$ and $(S_3, \bar{y}, S_4)$ are disjoint.
- $X_{\bullet}(t) \doteq r$ where $r \neq X_{\bullet}(t')$ contains term, context or sequence variables,
- $X_{\bullet}(C_1) \doteq C_2$ where $C_2 \neq X_{\bullet}(C_3)$ and C_2 is not strict.

A constraint is *solved*, if it is either true or a non-empty quantifier-free disjunction of solved conjunctions. A constraint is *partially solved*, if it is either true or a non-empty quantifier-free disjunction of partially solved conjunctions.

5 Solver

In this section we present a constraint solver. It is based on rules, transforming a constraint in *disjunctive normal form* (DNF) into a constraint in DNF. We say a constraint is in DNF, if it has a form $\mathcal{K}_1 \vee \cdots \vee \mathcal{K}_n$, where $\mathcal{K}$'s are conjunctions of true, false, and primitive constraints. The number of solver rules is not small (as it is usual for such kind of solvers, cf., e.g., [4, 5]). To make their comprehension easier, we group them so that similar ones are collected together in subsections. Within each subsection, for better readability, the rule groups are put between horizontal lines.

Before going into the details, we introduce a more conventional way of writing expressions, some kind of syntactic sugar, that should make reading easier. We write $F_1 \doteq F_2$ instead of $F_1() \doteq F_2()$, $X_{\bullet} \doteq C$ instead of $X_{\bullet}(\bullet) \doteq C$. The symmetric closure of $\doteq$ is denoted by $\simeq$.

5.1 Logical Rules

The logical rules perform logical transformations of the constraints and have to be applied in constraints, at any depth modulo associativity and commutativity of disjunction and conjunction. $\mathbf{F}$ stands for any formula. We denote the whole set of rules by Log.

$\mathbf{F} \wedge \mathbf{F} \rightsquigarrow \mathbf{F}$	$\text{true} \wedge \mathbf{F} \rightsquigarrow \mathbf{F}$	$\text{false} \wedge \mathbf{F} \rightsquigarrow \text{false}$	$S \simeq S \rightsquigarrow \text{true}$
$\mathbf{F} \vee \mathbf{F} \rightsquigarrow \mathbf{F}$	$\text{true} \vee \mathbf{F} \rightsquigarrow \text{true}$	$\text{false} \vee \mathbf{F} \rightsquigarrow \mathbf{F}$	$C \simeq C \rightsquigarrow \text{true}$

5.2 Failure Rules

In the second group there are rules for failure detection. The first two rules detect function symbol clash:

(F1)	$f_1(S_1) \simeq f_2(S_2) \rightsquigarrow$ false, if $f_1 \neq f_2$.
(F2)	$f_1(S_1, C_1, S_2) \simeq f_2(S_3, C_2, S_4) \rightsquigarrow$ false, if $f_1 \neq f_2$.

The next three rules perform occurrence check. Peculiarity of this operation for our language is that the variable occurrence into a term/context does not always leads to failure. For instance, the equation $x \doteq X_{\bullet}(x)$, where the variable x occurs in $X_{\bullet}(x)$, still has a solution $\{X_{\bullet} \mapsto \bullet\}$. Therefore, the occurrence check should fail on equations of the form $var \doteq nonvar$ only if no instance of the non-variable expression $nonvar$ can become the variable var. To achieve this, the rules below require the non-variable terms to contain F (the first two rules) and t (the third rule), which can not be erased by a substitution application:

(F3)	$x \simeq C[F(S)] \rightsquigarrow$ false, if $x \in var(C[F(S)])$.
(F4)	$X_{\bullet} \simeq C_1[F(C_2)] \rightsquigarrow$ false, if $X_{\bullet} \in var(C_1[F(C_2)])$.
(F5)	$\overline{x} \simeq (S_1, t, S_2) \rightsquigarrow$ false, if $\overline{x} \in var((S_1, t, S_2))$.

Further, we have two more rules which lead to failure in the case when the hole is unified with a context whose all possible instances are nontrivial contexts (guaranteed by the presence of F), and when the empty sequence is attempted to match to an inherently nonempty sequence (guaranteed by t):

(F6) $\bullet \simeq C_1[F(C_2)] \rightsquigarrow$ false.	(F7) $() \simeq (S_1, t, S_2) \rightsquigarrow$ false.

We denote this set of rules (F1)–(F7) by Fail.

5.3 Deletion Rules

There are five rules which delete identical terms, sequence variables or context variables from both sides of an equation. They are more or less self-explanatory. Just note that under unordered head, we delete an arbitrary occurrence of a term (that is not a sequence variable).

(Del1) $X_\bullet(t_1) \simeq X_\bullet(t_2) \wedge \leadsto t_1 \doteq t_2.$

(Del2) $X_\bullet(C_1) \simeq X_\bullet(C_2) \leadsto C_1 \doteq C_2.$

(Del3) $f_u(S_1, S, S_2) \simeq f_u(S_3, S, S_4) \leadsto f_u(S_1, S_2) \doteq f_u(S_3, S_4).$

(Del4) $(\overline{x}, S_1) \simeq (\overline{x}, S_2) \leadsto S_1 \doteq S_2.$

(Del5) $\overline{x} \simeq (S_1, \overline{x}, S_2) \leadsto S_1 \doteq (\,) \wedge S_2 \doteq (\,),$

In the last rule $\tilde{s}_1$ is not the empty sequence.

We denote the set of rules (Del1)–(Del5) by Del.

5.4 Decomposition Rules

Like the membership rules, each of the decomposition rules operates on a conjunction of constraint literals and gives back either a conjunction again, or a disjunction of conjunctions. These rules should be applied to disjuncts of constraints in DNF, to preserve the DNF structure.

(D1) $f_o(S_1) \simeq f_o(S_2) \leadsto S_1 \doteq S_2.$

(D2) $f_u(S_1) \simeq f_u(S_2) \wedge \mathcal{K} \leadsto \bigvee_{S' \in perm(S_2)} (S_1 \doteq S' \wedge \mathcal{K}),$

where S_2 is a sequence of terms, S_1 and S_2 are disjoint.

(D3) $(t_1, S_1) \simeq (t_2, S_2) \leadsto t_1 \doteq t_2 \wedge S_1 \doteq S_2,$

where $S_1 \neq (\,)$ or $S_2 \neq (\,)$.

(D4) $f(S_1, C_1, S_2) \simeq f(S_3, C_2, S_4) \leadsto f(S_1, S_2) \doteq f(S_3, S_4) \wedge C_1 \doteq C_2.$

We denote the set of rules (D1)–(D4) by Dec.

5.5 Variable Elimination Rules

This set of rules eliminate variables from the given constraint, keeping only a single equation for them. The first four rules replace a variable with the corresponding expression, provided that the variable does not occur in the expression:

$$\text{(E1)} \quad x \simeq t \wedge \mathcal{K} \rightsquigarrow x \doteq t \wedge \mathcal{K}\theta,$$

where $x \notin var(t)$, $x \in var(\mathcal{K})$ and $\theta = \{x \mapsto t\}$. If t is a variable then in addition it is required that $t \in var(\mathcal{K})$.

$$\text{(E2)} \quad \overline{x} \simeq S \wedge \mathcal{K} \rightsquigarrow \overline{x} \doteq S \wedge \mathcal{K}\theta,$$

where $\overline{x} \notin var(S)$, $\overline{x} \in var(\mathcal{K})$, and $\theta = \{\overline{x} \mapsto S\}$. If $S = \overline{y}$ for some $\overline{y}$, then in addition it is required that $\overline{y} \in var(\mathcal{K})$.

$$\text{(E3)} \quad X_{\bullet} \simeq C \wedge \mathcal{K} \rightsquigarrow X_{\bullet} \doteq C \wedge \mathcal{K}\theta,$$

where $X_{\bullet} \notin var(C)$, $X_{\bullet} \in var(\mathcal{K})$, and $\theta = \{X_{\bullet} \mapsto C\}$. If C has the form $Y_{\bullet}(\bullet)$, then in addition it is required that $Y_{\bullet} \in var(\mathcal{K})$.

$$\text{(E4)} \quad X \simeq F \wedge \mathcal{K} \rightsquigarrow X \doteq F \wedge \mathcal{K}\theta,$$

where $X \neq F$, $X \in var(\mathcal{K})$, and $\theta = \{X \mapsto F\}$. If F is a function variable, then in addition it is required that $F \in var(\mathcal{K})$.

The rules (E5) and (E6) for sequence variable elimination assign to a variable an initial part of the sequence in the other side of the selected equation. The sequence has to be a sequence of terms in (E5). In (E6), only a split of the prefix of the sequence is relevant. The rest is blocked by the term t due to occurrence check: No instantiation of $\overline{x}$ can contain it.

$$\text{(E5)} \quad (\overline{x}, S_1) \simeq S_2 \wedge \mathcal{K} \rightsquigarrow \bigvee_{S_2=(S',S'')} \left(\overline{x} \doteq S' \wedge S_1\theta \doteq S'' \wedge \mathcal{K}\theta \right)$$

where S_2 is a sequence of terms, $\overline{x} \notin var(S_2)$, $\theta = \{\overline{x} \mapsto S'\}$, and $S_1 \neq (\,)$.

$$\text{(E6)} \quad (\overline{x}, S_1) \simeq (S, t, S_2) \wedge \mathcal{K} \rightsquigarrow$$
$$\bigvee_{S=(S',S'')} \left(\overline{x} \doteq S' \wedge S_1\theta \doteq (S'', t, S_2)\theta \wedge \mathcal{K}\theta \right)$$

where S is a term sequence, $\overline{x} \notin var(S)$, $\overline{x} \in var(t)$, $\theta = \{\overline{x} \mapsto S'\}$, and $S_1 \neq (\,)$.

The rules (E7) and (E8) below can be seen as counterparts of (E5). In the rule (E8) we need conservative decomposition of contexts. Before giving those rules, we define the notion of conservativity.

We will speak about the *main path* of a context as the sequence of symbols (path) in its tree representation from the root to the hole. For instance, the main path in the context $f(X_{\bullet 1}(a), X(X_{\bullet 2}(b), g(\bullet)), \overline{x})$ is fXg, and in $f(X_{\bullet 1}(a), X(X_{\bullet 2}(b), X_{\bullet 3}(\bullet)), \overline{x}) - fXX_{\bullet 3}$. A context is called *strict* if its main path does not contain context variables. For instance, the context $f(X_{\bullet 1}(a), X(X_{\bullet 2}(b), g(\bullet)), \overline{x})$ is strict, while $f(X_{\bullet 1}(a), X(X_{\bullet 2}(b), X_{\bullet 3}(\bullet)), \overline{x})$ is not, because $X_{\bullet 3}$ is in its main path $fXX_{\bullet 3}$. We say that a context C is *decomposed* in two contexts C_1 and C_2 if $C = C_1[C_2]$.

We say that a context C is *conservative*, if for any instance $C\rho$ of C and for any decomposition $D_1[D_2]$ of $C\rho$ there exists a decomposition $C_1[C_2]$ of C such that $D_1 = C_1\rho$ and $D_2 = C_2\rho$. Strict contexts satisfy this property. Non-strict contexts violate it, as the following example shows: The context $C = X_\bullet(\bullet)$ has two decompositions into $C_1[C_2]$ with $C_1 = \bullet$, $C_2 = X_\bullet(\bullet)$ and $C_1 = X_\bullet(\bullet)$, $C_2 = \bullet$. Let $\rho = \{X_\bullet \mapsto f(g(\bullet))\}$. Then $C\rho = f(g(\bullet))$. One of its decomposition with $D_1 = f(\bullet)$, $D_2 = g(\bullet)$ is not an instance of any of the decompositions of C.

The rules (E7) and (E8) are formulated now as follows:

$$(\text{E7}) \quad X_\bullet(t_1) \simeq t_2 \wedge \mathcal{K} \rightsquigarrow \vee_{t_2=C[t]}\left(X_\bullet \doteq C \wedge t_1\theta \doteq t \wedge \mathcal{K}\theta\right),$$

where t_2 does not contain term, sequence, and context variables, $t_1 \neq \bullet$, and $\theta = \{X_\bullet \mapsto C\}$.

$$(\text{E8}) \quad X_\bullet(C_1) \simeq C_2 \wedge \mathcal{K} \rightsquigarrow \vee_{C_2=C[C']}\left(X_\bullet \doteq C \wedge C_1\theta \doteq C'\theta \wedge \mathcal{K}\theta\right),$$

where C_2 is strict, $X_\bullet \notin var(C)$, $C_1 \neq \bullet$, and $\theta = \{X_\bullet \mapsto C\}$.

Finally, there are two rules for function variable elimination. Their behavior is standard:

$$(\text{E9}) \quad X(S_1) \simeq F(S_2) \wedge \mathcal{K} \rightsquigarrow X \doteq F \wedge F(S_1)\theta \doteq F(S_2)\theta \wedge \mathcal{K}\theta.$$

where $X \neq F$, $\theta = \{X \mapsto F\}$, and $S_1 \neq ()$ or $S_2 \neq ()$.

$$(\text{E10}) \quad X(S_1) \simeq X(S_2) \wedge \mathcal{K} \rightsquigarrow \bigvee_{f\in\mathcal{F}}\left(X \doteq f \wedge f(S_1)\theta \doteq f(S_2)\theta \wedge \mathcal{K}\theta\right),$$

where $\theta = \{X \mapsto f\}$, and $S_1 \neq S_2$.

We denote the set of rules (E1)–(E10) by Elim.

The constraint solver rewrites a constraint with respect to the rules specified in this section into a constraint in partially solved form. First, we define how rewriting is done in a single step:

$$\text{step} := \text{first}(\text{Log, Fail, Del, Dec, Elim}).$$

When step is applied to a constraint, it will transforms the constraint by the *first* successful rule from the sets Log, Fail, Del, Dec, and Elim. If none of the rules apply, then the constraint is said to be in a *normal form* with respect to step.

The constraint solving method implements the strategy solve which is defined as a repeatedly application of the step:

$$\text{solve} := \text{NF}(\text{step}).$$

That means, step is applied to a constraint repeatedly as long as possible.

It remains to show that this definition yields an algorithm, which amounts to proving that a normal form is reached by $\mathsf{NF}(\mathsf{step})$ for any constraint C.

6 Properties of the Constraint Solver

In this section, we present theorems and lemmata which demonstrate that the constraint solver is terminated, sound and partially complete. The proofs are omitted and can be easily obtained from the proofs of the similar theorems and lemmata given in [6].

The solver halts for any input constraint and a normal form is reached.

Theorem 1 solve *terminates on any input constraint.*

Here we state that the solver reduces a constraint to its equivalent constraint.

Lemma 1 *If* $\mathsf{step}(C) = D$, *then* $\mathfrak{I} \models \forall \left(C \Leftrightarrow \overline{\exists}_{var(C)} D \right)$ *for all intended structures* $\mathfrak{I}$.

Theorem 2 *If* $\mathsf{solve}(C) = D$, *then* $\mathfrak{I} \models \forall \left(C \Leftrightarrow \overline{\exists}_{var(C)} D \right)$ *for all intended structures* $\mathfrak{I}$, *and* D *is either partially solved or the* false *constraint.*

Theorem 3 *If the constraint* D *is solved, then* $\mathfrak{I} \models \exists D$ *for all intended structures* $\mathfrak{I}$.

7 Well-Moded Constraints

A sequence of primitive constraints $\mathbf{c}_1, \ldots, \mathbf{c}_n$ is *well-moded* if the following conditions are satisfied:

1. If for some $1 \leq i \leq n$, $\mathbf{c}_i$ is $t_1 \doteq t_2$, then $var(t_1) \subseteq \bigcup_{j=1}^{i-1} var(\mathbf{c}_j)$ or $var(t_2) \subseteq \bigcup_{j=1}^{i-1} var(\mathbf{c}_j)$.
2. If for some $1 \leq i \leq n$, $\mathbf{c}_i$ is $C_1 \doteq C_2$, then $var(C_1) \subseteq \bigcup_{j=1}^{i-1} var(\mathbf{c}_j)$ or $var(C_2) \subseteq \bigcup_{j=1}^{i-1} var(\mathbf{c}_j)$.

A conjunction of primitive constraints $\mathcal{K}$ is well-moded if there exists a sequence of primitive constraints $\mathbf{c}_1, \ldots, \mathbf{c}_n$ which is well-moded and $\mathcal{K} = \bigwedge_{i=1}^{n} \mathbf{c}_i$ modulo associativity and commutativity of $\wedge$. A constraint $C = \mathcal{K}_1 \vee \cdots \vee \mathcal{K}_n$ is well-moded if each $\mathcal{K}_i$, $1 \leq i \leq n$, is well-moded.

The following Theorem states, that, the solver brings any well-moded constraints to a solved form or to false.

Lemma 2 *Let* C *be a well-moded constraint and* $\mathsf{step}(C) = C'$, *then* C' *is either well-moded,* true *or* false.

Theorem 4 *Let C be a well-moded constraint and* solve(C) $= C'$, *where* $C' \neq$ false. *Then C' is solved.*

Acknowledgments Besik Dundua has been partially supported by the Shota Rustaveli National Science Foundation under the grants FR/325/4-120/14, YS/10/11-811/15 and YS15_2.1.2_70.

References

1. Bojanczyk, M., Walukiewicz, I.: Forest algebras. In: Flum, J., Grädel, E., Wilke, T. (eds.) Logic and Automata. Texts in Logic and Games, vol. 2, pp. 107–132. Amsterdam University Press (2008)
2. Boley, H.: A Tight, Practical Integration of Relations and Functions. Lecture Notes in Computer Science, vol. 1712. Springer, Berlin (1999)
3. Chasseur, E., Deville, Y.: Logic program schemas, constraints, and semi-unification. In: Fuchs, N.E. (ed.) LOPSTR. Lecture Notes in Computer Science, vol. 1463, pp. 69–89. Springer, Berlin (1997)
4. Comon, H.: Completion of rewrite systems with membership constraints. Part II: constraint solving. J. Symb. Comput. **25**(4), 421–453 (1998)
5. Dovier, A., Piazza, C., Pontelli, E., Rossi, G.: Sets and constraint logic programming. ACM Trans. Program. Lang. Syst. **22**(5), 861–931 (2000)
6. Dundua, B.: Programming with Sequence and Context Variables:Foundations and Applications. Ph.D. thesis, Universidade do Porto (2014)
7. Dundua, B., Florido, M., Kutsia, T., Marin, M.: Constraint logic programming for hedges: A semantic reconstruction. In: Codish, M., Sumii, E. (eds.) Functional and Logic Programming—12th International Symposium, FLOPS 2014, Kanazawa, Japan, June 4—6, 2014. Proceedings. Lecture Notes in Computer Science, vol. 8475, pp. 285–301. Springer, Switzerland (2014)
8. Dundua, B., Kutsia, T., Marin, M.: Strategies in Pρlog. In: Fernández, M. (ed.) WRS, EPTCS, vol. 15, pp. 32–43 (2009)
9. ISO/IEC. Information technology—Common Logic (CL): a framework for a family of logic-based languages. International Standard ISO/IEC 24707 (2007). http://standards.iso.org/ittf/PubliclyAvailableStandards/c039175_ISO_IEC_24707_2007(E).zip
10. Jacquemard, F., Rusinowitch, M.: Closure of hedge-automata languages by hedge rewriting. In: Voronkov, A. (ed.) RTA. LNCS, vol. 5117, pp. 157–171. Springer, Berlin (2008)
11. Kutsia, T.: Solving and Proving in Equational Theories with Sequence Variables and Flexible Arity Symbols. RISC report Series 02–09, Research Institute for Symbolic Computation (RISC), University of Linz, Schloss Hagenberg, 4232 Hagenberg, Austria, May 2002. Ph.D. thesis
12. Kutsia, T.: Theorem proving with sequence variables and flexible arity symbols. In: Baaz, M., Voronkov, A. (eds.) LPAR. Lecture Notes in Computer Science, vol. 2514, pp. 278–291. Springer, Berlin (2002)
13. Kutsia, T.: Solving equations with sequence variables and sequence functions. J. Symb. Comput. **42**(3), 352–388 (2007)
14. Kutsia, T., Buchberger, B.: Predicate logic with sequence variables and sequence function symbols. In: Asperti, A., Bancerek, G., Trybulec, A. (eds.) MKM. Lecture Notes in Computer Science, vol. 3119, pp. 205–219. Springer, Berlin (2004)
15. Kutsia, T., Marin, M.: Matching with regular constraints. In: Sutcliffe, G., Voronkov, A. (eds.) LPAR. Lecture Notes in Computer Science, vol. 3835, pp. 215–229. Springer, Berlin (2005)
16. Kutsia, T., Marin, M.: Solving, reasoning, and programming in common logic. In: SYNASC, pp. 119–126. IEEE Computer Society (2012)
17. Levy, J.: Linear second-order unification. In: Ganzinger, H. (ed.) RTA. Lecture Notes in Computer Science, vol. 1103, pp. 332–346. Springer, Berlin (1996)

18. Libkin, L.: Logics for unranked trees: an overview. Logical Methods Comput. Sci. **2**(3) (2006)
19. Makanin, G.S.: The problem of solvability of equations in a free semigroup. Math. USSR-Sb. **32**(2), 129, 147–236 (1977)
20. Menzel, C.: Knowledge representation, the world wide web, and the evolution of logic. Synthese **182**(2), 269–295 (2011)
21. Schmidt-Schauß, M., Stuber, J.: The complexity of linear and stratified context matching problems. Theor. Comput. Syst. **37**(6), 717–740 (2004)
22. Wand, M.: Complete type inference for simple objects. In: LICS, pp. 37–44. IEEE Computer Society (1987)
23. Wolfram, S.: The Mathematica Book. Wolfram-Media, 5th edn. (2003)

Hyperpath Centers

Mehmet Ali Balcı, Sibel Paşalı Atmaca and Ömer Akgüller

Abstract Path centers play an important role for several kinds of networks. In this study we present a novel approach to find a path center for a spatial data set modelled as a hypergraph. For this purpose, we first present a simple graph representation of a hypergraph, then determine the central path by eccentricities.

1 Introduction

The notion of the centrality in a network has the key role to model real world applications typically involve problems of finding optimal locations in a network and has been studied by many researchers. In several studies such as [8, 13, 14], finding the optimal location of a facility is studied by considering the central paths. Examples of such facilities railroad lines, highways, pipelines, transit routes, are presented in Slater [17]. Also the path center has clear applications to social network theory [7], and biochemical networks. In Junker et al. [9], it is shown that central elements of biological networks have been found to be essential for viability.

Hypergraphs are generalization of the simple graphs that more efficiently model complex systems since they are one of the most general mathematical structures for representing relationships. They are applied in several areas as image processing [20], cybernetics [23], to machine learning [21, 22]. The concept of hyperpath and its applicaitons can also be found in [10, 11, 19]. Hypergraphs are also essential tools in modeling spatial data sets since they provide more benefits rather than simple graphs. One of the major advantages is that the hypergraph model allows us to effectively represent important relations among data points in data structure on which computationally efficient clustering approaches can be used to find clusters of related data points [3].

M.A. Balcı (✉) · S. Paşalı Atmaca · Ö. Akgüller
Faculty of Science, Department of Mathematics, Mugla Sitki Kocman University,
48000 Merkez, Muğla, Turkey
e-mail: mehmetalibalci@mu.edu.dr

© Springer International Publishing Switzerland 2016   129
T.B. Nguyen et al. (eds.), *Advanced Computational Methods
for Knowledge Engineering*, Advances in Intelligent Systems
and Computing 453, DOI 10.1007/978-3-319-38884-7_10

In this study we briefly studied the central path for a hypergraph. There are several ways to represent a hypergraph as a simple graph [1]. The most common ones are the bipartite and path graphs. Even though the hypergraphs can model a network more efficiently, calculation of some basic concepts such as paths or hypertrees can cost much more effort. To deal with this problem, we present a new simple graph representation of a hypergraph called neighborhood graph. To the best of our knowledge, this kind of simple graph representation is the first in the literature. Since it is directly depend on the connectivity of each vertices, the hypergraph analogues of the concepts such as paths, centralities, and vulnerability measures can be studied more efficiently. In Sect. 2 we first give some basic definitions to obtain our goal. In Sect. 3, we first introduce the neighborhood graph of a hypergraph then a method to find central hyperpaths. We also give the results of our method for the randomly generated 30 spatial points by using the Delaunay triangulation and its hypergraph representation.

2 Preliminaries

An undirected graph G is the tuples (V, E), where V is the set of vertices and E is the set of edges. Each elements of E is an unordered pair of vertices for an undirected graph G. For $V' \subset V$ and $E' \subset E$, the tuples $G' = (V', E')$ is called the subgraph of V. For $u, v \in V$, a distance $d(u, v)$ in $G = (V, E)$ is the shortest path, that is the shortest sequence of edges which connect a sequence of vertices, between them. For any vertex v in G and the subgraph G', the distance $d(v, G')$ is the minimum distance from v to a vertex in G'. The eccentricity of G' is the distance to a vertex farthest from G' and shown as $e(G')$. Therefore, for $v \in G$, $e(G') = \max\{d(v, G')\}$.

Definition 1 A path P is a path center of G if P has minimum eccentricity and has minimum length among such paths [2].

For the tree in Fig. 1, paths $v_2 v_3 v_4 v_5$ and $v_3 v_4 v_5 v_6$ have eccentricity 3 and 2, respectively. The central path is $v_4 v_5 v_6$ with eccentricity 2.

An algorithm for finding the path center of a simple graph is briefly introduced in [4]. Here Latin and Floyd algorithms are both used. The adjacency matrix of the graph and a text file which hold the paths can be found in the graph that is used. The complexity of the algorithm is $O(n^3)$. So it is much more useful in the solution of such problem models rather than other solution methods.

A hypergraph $\mathcal{H} = (\mathcal{V}, \mathcal{E})$ consists of a non-empty vertex set $\mathcal{V} = \{v_i | i = 1, \dots, p\}$ and a non-empty family of hyperedges or hyperarcs $\mathcal{E} = \{e_j | j = 1, \dots, q\}$.

In a hypergraph, a hyperpath is defined as follows:

Definition 2 A hyperpath between two vertices u and v is a sequence of hyperedges $\{e_1, e_2, \dots, e_m\}$ such that $u \in e_1$, $v \in e_m$, and $e_i \cap e_{i+1} \neq \emptyset$ for all $i = 1, 2, \dots, m-1$. A hyperpath is simple if non-adjacent hyperedges in the path are non-overlapping [6].

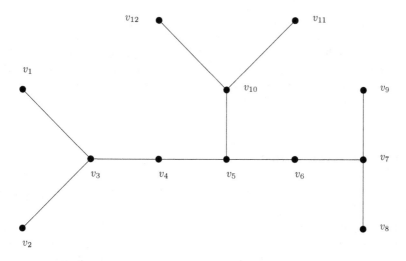

Fig. 1 A tree to illustrate path center

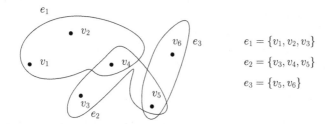

$$e_1 = \{v_1, v_2, v_3\}$$
$$e_2 = \{v_3, v_4, v_5\}$$
$$e_3 = \{v_5, v_6\}$$

Fig. 2 A Hypergraph $\mathcal{H}$ with $\mathcal{V} = \{v_1, v_2, v_3, v_4, v_5, v_6\}$ and $\mathcal{E} = \{e_1, e_2, e_3\}$

For the hypergraph in Fig. 2, the paths $\{v_1 = e_1, e_2, e_3 = v_6\}$ and $\{v_2 = e_1, e_2, e_3 = v_6\}$ yield the same sequence but have different initial vertices.

The spatial data sets are the sets of data or an information system that identify the location of features that could be the city locations of the global route map, the locations of the natural or constructed features, or even the locations of modules in an integrated circuit chip. In mathematics, the Delaunay triangulation for a set $\mathcal{P}$ of points in the plane is a triangulation $D(\mathcal{P})$ such that no point in $\mathcal{P}$ is inside the circum circle of any triangle in $D(\mathcal{P})$. This kind of triangulation of spatial data sets has varieties of applications from Geographic Information systems [12, 16], to remote sensing [15]. For the given set of data points $\mathcal{P} = \{p_1, p_2, \ldots, p_n\}$ the Voronoi region of is the locus of points which have as a Euclidean nearest neighbor; i.e., $\{x \in \mathbb{R}^2 | \forall j \neq i, d_E(x, p_i) \leq d_E(x, p_j)\}$, where d_E is the Euclidean distance. The regions are convex polygons, and their interiors are disjoint [5]. The Delaunay triangulation $D(\mathcal{P})$ of $\mathcal{P}$ is a planar graph embedding such that the nodes of $D(\mathcal{P})$ consist of the data points of $\mathcal{P}$, and two nodes p_i, p_j are joined by an edge if the boundaries of the corresponding Voronoi regions share a line segment. By considering the Voronoi regions for 2D spatial data sets, there are several different algorithms for computing the Delaunay triangulation. In [18]; Dwyers divide and conquer algorithm, Fortunes sweepline algorithm, several versions of the incremental algorithms, an algorithm that incrementally adds a correct Delaunay triangle adjacent to a current triangle in a manner similar to gift wrapping algorithms for convex hulls, and Barbers convex hull based algorithm are briefly examined.

3   Central Hyperpaths

In this section, we give the algorithmic procedure that to obtain central hyperpaths in any hypergraph. Even this procedure can be executed for any hypergraph, we apply our method to the hypergraph representation of the Delaunay Triangulation of a spatial data set. Since the main goal is to obtain hyperpaths, we first introduce a simple graph representation that is depending on the connectivity of the hypergraph.

Definition 3 Let $\mathcal{H} = (\mathcal{V}, \mathcal{E})$ be a hypergraph with $|\mathcal{V}| = n$. The matrix $B_{e_k} = [a_{ij}]_{n \times n}$ whose entries are

$$a_{ij} = \begin{cases} 1, & \{v_i, v_j\} \subset e_k \text{ for } i \neq j \\ 0, & \text{otherwise} \end{cases}$$

is called the neighborhood matrix of the edge $e_k \in \mathcal{E}$.

Definition 4 Let $\mathcal{H} = (\mathcal{V}, \mathcal{E})$ be a hypergraph with $|\mathcal{V}| = n$ and $|\mathcal{E}| = m$. The simple graph $G_{\mathcal{H}}$ with the adjacency matrix

$$A_{G_{\mathcal{H}}} = B_{e_1} \oplus B_{e_2} \oplus \ldots \oplus B_{e_m},$$

where $\oplus$ is the element-wise Boolean sum of neighborhood matrices is called the neighborhood graph of the hypergraph $\mathcal{H}$.

We may remark that the neighborhood graph of any simple graph is itself. Also any non-cyclical path between two vertices in neighborhood graph corresponds a non-overlapping hyperpath in $\mathcal{H}$.

To determine a central hyperpath in a given hypergraph $\mathcal{H}$, we execute the following procedure:

Input:	A hypergraph $\mathcal{H} = (\mathcal{V}, \mathcal{E})$
Step 1:	Construct the neighborhood graph $G_{\mathcal{H}}$
Step 2:	Determine all non-cyclical path with length k between all vertices in $G_{\mathcal{H}}$
Step 3:	For each path P_i, calculate the eccentricities
Step 4:	List the P_i which have minimum eccentricity

Table 1 Algorithm for step 1

Input: $\mathcal{H} = (\mathcal{V}, \mathcal{E})$
for $k = 1$ **to** $\|\mathcal{E}\|$
for $i = 1$ **to** $\|\mathcal{V}\|$
for $j = 1$ **to** $\|\mathcal{V}\|$
if $v_i \wedge v_j \in e_k$ **then** $B_k(i,j) = 1$ **else** $B_k(i,j) = 0$
end if
end for
end for
end for
$Adj_{\mathcal{H}} = \bigoplus\limits_{k=1}^{\|\mathcal{E}\|} B_k$
$G_{\mathcal{H}} \leftarrow$ the graph with the adjacency matrix $Adj_{\mathcal{H}}$
Output: The neighborhood graph $G_{\mathcal{H}}$ of $\mathcal{H}$

The construction of the neighborhood graph $G_{\mathcal{H}}$ is given algorithmically in Table 1. Finding all paths with k length in $G_{\mathcal{H}}$ can be executed in the time complexity $O(|\mathcal{V}|^2|E_{G_{\mathcal{H}}}|)$, where $|E_{G_{\mathcal{H}}}|$ is the cardinality of the edge set of $G_{\mathcal{H}}$. In the real world applications, $|E_{G_{\mathcal{H}}}|$ is expected to be greater and equal than $|\mathcal{E}|$, hence the purposed algorithm has the time complexity $O(|\mathcal{V}|^3)$ in the worst case. In this study we used the algorithm that is introduced in [4] for all possible k length paths in $G_{\mathcal{H}}$. For any vertex v in $G_{\mathcal{H}}$ and the path P_i, the distance $d(v, P_i)$ is equivalent to the Hausdorff distance $d_H(v, \mathcal{H}')$, where $\mathcal{H}'$ is the sub-hypergraph that P_i yields in $\mathcal{H}$. Therefore, the neighborhood graph is pretty useful to determine central paths.

3.1 Application to a 2D Spatial Data Set

Let us consider the hypergraph $\mathcal{H} = (\mathcal{V}, \mathcal{E})$ that is obtained from the Delaunay Triangulation of randomly generated 30 points in $\mathbb{R}^2$. Randomly generated points and their Delaunay Triangulation is given in Fig. 3.

$$
\begin{cases}
(-8.68431, -19.4597) & (-12.0276, -13.777) & (17.4263, -12.4599) \\
(7.36626, -16.1525) & (-13.6486, -19.299) & (-16.1117, 8.16653) \\
(-5.44457, -13.5817) & (-10.5417, -18.0569) & (-6.49097, -3.18305) \\
(16.1358, -17.3538) & (-18.7413, -0.719729) & (-12.2643, -14.7403) \\
(3.78991, -2.05521) & (3.81862, -8.40079) & (10.172, -18.0924) \\
(18.4567, -2.525) & (-8.76905, -12.7824) & (-5.3373, 4.11666) \\
(18.9651, -4.26288) & (-19.2284, 11.2134) & (3.1219, -7.02531) \\
(12.2673, 6.68884) & (-4.14705, -0.892884) & (12.8148, -13.9107) \\
(-7.16532, 11.5893) & (-9.11853, -7.90572) & (-1.49457, -6.06677) \\
(-16.6969, 17.0202) & (-0.163445, 13.8095) & (12.851, -7.69346)
\end{cases}
$$

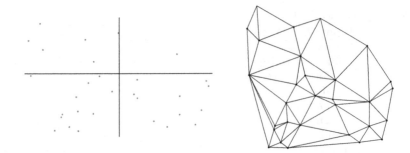

Fig. 3 Left side is the 2D spatial data randomly generated between [−30, 30] and right side is the Delaunay Triangulation

Each hyperedge $e_k \in \mathcal{E}$, connects vertices of a triangle in the Delaunay Triangulation of the data points. Hence, it involves three data points connected. To consider the low ordered relationship among the spatial data points, we discard the hyperedge corresponding the triangle that has the total length of three edges larger than other ones.

Afterwards the obtained hypergraph with $|\mathcal{E}| = 40$, we can construct the corresponding neighborhood graph, then obtain the central path by using the procedure given in Sect. 3. The obtained neighborhood graph is given in Fig. 4.

The algorithm determines 429 distinct path with 4 vertices in $G_{\mathcal{H}}$. For the different values of k up to 30 can also be analyzed. By considering the distance matrix, it is possible to determine paths with minimum eccentricities. These paths are given in Table 2, and the three of them are visualized on the neighborhood graph of $G_{\mathcal{H}}$ and Delaunay Graph of the spatial data points (Fig. 5).

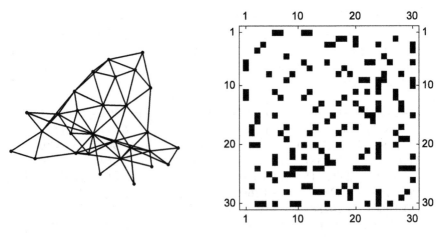

Fig. 4 The neighborhood graph of the hypergraph obtained from the Delaunay Triangulation and its sparse adjacency matrix

Table 2 Central paths with minimum eccentricities in $G_\mathcal{H}$

$(v_1, v_6, v_{11}, v_{20}, v_{17})$	$(v_1, v_6, v_{11}, v_{22}, v_9)$	$(v_1, v_6, v_{11}, v_{24}, v_{28})$	$(v_1, v_6, v_{22}, v_9, v_8)$
$(v_1, v_{11}, v_{24}, v_{28}, v_{19})$	$(v_2, v_{17}, v_8, v_9, v_{22})$	$(v_2, v_{17}, v_8, v_9, v_{30})$	$(v_2, v_{17}, v_{13}, v_{20}, v_3)$
$(v_2, v_{17}, v_{13}, v_{20}, v_{11})$	$(v_2, v_{17}, v_{20}, v_3, v_4)$	$(v_2, v_{17}, v_{20}, v_3, v_{15})$	$(v_2, v_{17}, v_{20}, v_3, v_{16})$
$(v_2, v_{17}, v_{20}, v_{11}, v_6)$	$(v_3, v_{20}, v_2, v_{17}, v_{18})$	$(v_3, v_{20}, v_2, v_{18}, v_{17})$	$(v_3, v_{20}, v_{17}, v_8, v_5)$
$(v_4, v_3, v_{20}, v_2, v_{17})$	$(v_4, v_3, v_{20}, v_{17}, v_8)$	$(v_4, v_3, v_{20}, v_{17}, v_{13})$	$(v_5, v_8, v_9, v_{22}, v_6)$
$(v_5, v_8, v_9, v_{22}, v_{11})$	$(v_6, v_1, v_7, v_{18}, v_{17})$	$(v_6, v_1, v_{11}, v_{22}, v_9)$	$(v_6, v_1, v_{11}, v_{24}, v_{28})$
$(v_6, v_{11}, v_{20}, v_{17}, v_8)$	$(v_6, v_{11}, v_{22}, v_9, v_{25})$	$(v_6, v_{11}, v_{24}, v_{28}, v_{19})$	$(v_7, v_1, v_6, v_{22}, v_9)$
$(v_7, v_1, v_{11}, v_{20}, v_{17})$	$(v_7, v_1, v_{11}, v_{24}, v_{28})$	$(v_8, v_9, v_{21}, v_{22}, v_{11})$	$(v_8, v_9, v_{22}, v_6, v_{23})$
$(v_8, v_9, v_{22}, v_6, v_{26})$	$(v_8, v_9, v_{22}, v_{11}, v_{12})$	$(v_8, v_9, v_{22}, v_{11}, v_{27})$	$(v_9, v_8, v_{17}, v_{20}, v_{11})$
$(v_{10}, v_{24}, v_3, v_{20}, v_{17})$	$(v_{11}, v_6, v_{22}, v_9, v_{25})$	$(v_{11}, v_6, v_{22}, v_{24}, v_{28})$	$(v_{11}, v_{20}, v_{13}, v_{28}, v_{19})$
$(v_{12}, v_1, v_{11}, v_{24}, v_{28})$	$(v_{12}, v_{11}, v_{22}, v_9, v_{25})$	$(v_{12}, v_{11}, v_{24}, v_{28}, v_{19})$	$(v_{15}, v_3, v_{20}, v_2, v_{17})$
$(v_{16}, v_3, v_{20}, v_2, v_{17})$	$(v_{17}, v_2, v_{20}, v_3, v_{24})$	$(v_{17}, v_2, v_{20}, v_3, v_{30})$	$(v_{17}, v_2, v_{20}, v_{11}, v_{22})$
$(v_{17}, v_8, v_9, v_{30}, v_{26})$	$(v_{17}, v_{20}, v_3, v_{15}, v_{23})$	$(v_{21}, v_9, v_{22}, v_{11}, v_{27})$	$(v_{21}, v_9, v_{24}, v_{11}, v_{22})$
$(v_{22}, v_6, v_{11}, v_{24}, v_{28})$	$(v_{23}, v_6, v_{11}, v_{24}, v_{28})$	$(v_{26}, v_6, v_{11}, v_{24}, v_{28})$	

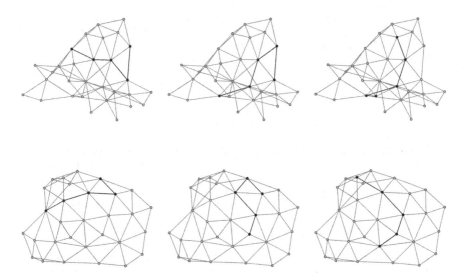

Fig. 5 The central hyperpaths end their representations. The upper graphs are the neighborhood graphs and the lowers are the Delaunay Graph representation of the spatial data points. The chosen hyperpaths are $(v_1, v_6, v_{11}, v_{20}, v_{17})$ $(v_6, v_1, v_{11}, v_{22}, v_9)$ and $(v_{21}, v_9, v_{22}, v_{11}, v_{27})$ respectively from left to the right

4 Conclusions

In this paper, we have presented a novel method which finds central hyperpaths in a hypergraph model. The discussed hypergraph is initially constructed from the Delaunay triangulation graph of the data set and can capture the relationships among sets of 2D data points. The hyperpaths playing central role in the hypergraph model are obtained by the minimum eccentricities among all possible paths in the presented neighborhood graph representation. This method is promising since it can also be extended to the data sets in 3D by considering the Delaunay Tetrahedralization, or even to the data sets in 4D where the time is the codimension.

References

1. Berge, C.: Graphs and Hypergraphs, vol. 7. North-holland Publishing Company, Amsterdam (1973)
2. Buckley, F., Harary, F.: Distance in Graphs. Addison Wesley Pub, California (1990)
3. Cherng, J.S., Lo, M.J.: A hypergraph based clustering algorithm for spatial data sets, data mining. In: Proceedings IEEE International Conference on ICDM 2001, pp. 83–90. IEEE (2001)
4. Dündar, P., Kılıç, E., Balcı, M.A.: Finding the path center of a communication network. Neural Netw. World 19(6), 725–733 (2009)
5. Estivill-Castro, V., Houle, M.E.: Robust distancebased clustering with applications to spatial data mining. Algorithmica 30(2), 216242 (2001)
6. Gao, J., Qing, Z., Wei, R., Ananthram, S., Ram, R., Amotz, B.N.: Dynamic shortest path algorithms for hypergraphs. In: 10th International Symposium on Modeling and Optimization in Mobile, Ad Hoc and Wireless Networks (WiOpt), pp. 238–245. IEEE (2012)
7. Hage, P., Harary, F.: Eccentricity and centrality in networks. Soc. Netw. 31, 17(1), 57–63 (1995)
8. Hedetniemi, S.M., Cockayne, E.J., Hedetniemi, S.T.: Linear algorithms for finding the Jordan center and path center of a tree. Transp. Sci. 15(2), 98–114 (1981)
9. Junker, B.H., Koschtzki, D., Schreiber, F.: Exploration of biological network centralities with CentiBiN. BMC Bioinf. 21, 7(1), 219 (2006)
10. Lozano, A., Storchi, G.: Shortest viable hyperpath in multimodal networks. Transp. Res. Part B: Methodol. 36(10), 853–874 (2002)
11. Marcotte, P., Nguyen, S.: Hyperpath formulations of traffic assignment problems. In: Equilibrium and Advanced Transportation Modelling, pp. 175–200. Springer, New York (1998)
12. Miller, H.J., Wentz, E.A.: Representation and spatial analysis in geographic information systems. Ann. Assoc. Am. Geogr. 93(3), 574–594 (2003)
13. Minieka, E.: The optimal location of a path or tree in a tree network. Networks 15(3), 309–21 (1985)
14. Morgan, C.A., Slater, P.J.: A linear algorithm for a core of a tree. J. Algorithms 30, 1(3), 247–58 (1980)
15. Riano, D., Meier, E., Allgöwer, B., Chuvieco, E., Ustin, S.L.: Modeling airborne laser scanning data for the spatial generation of critical forest parameters in fire behavior modeling. Remote Sens. Environ. 86(2), 177–186 (2003)
16. Shi, W., Pang, M.Y.C.: Development of Voronoi-based cellular automata-an integrated dynamic model for Geographical Information Systems. Int. J. Geogr. Inf. Sci. 14(5), 455–474 (2000)
17. Slater, P.J.: Locating central paths in a graph. Transp. Sci. 16(1), 1–8 (1982)
18. Su, P., Drysdale, R.L.S.: A comparison of sequential Delaunay triangulation algorithms. In: Proceedings of the eleventh annual symposium on Computational geometry, pp. 61–70. ACM (1995)
19. Underwood, P.C., Chamarthi, B., Williams, J.S., Sun, B., Vaidya, A., Raby, B.A., Lasky-Su, J., Hopkins, P.N., Adler, G.K., Williams, G.H.: Replication and meta-analysis of the gene-environment interaction between body mass index and the interleukin-6 promoter polymorphism with higher insulin resistance. Metabolism 61(5), 667–671 (2012)

20. Yu, J., Tao, D., Wang, M.: Adaptive hypergraph learning and its application in image classification. IEEE Trans. Image Process. **21**(7), 3262–3272 (2012)
21. Yu, J., Tao, D., Li, J., Cheng, J.: Semantic preserving distance metric learning and applications. Inf. Sci. **281**, 674–686 (2014)
22. Zhang, H., Yu, J., Wang, M., Liu, Y.: Semi-supervised distance metric learning based on local linear regression for data clustering. Neurocomputing **93**, 100–105 (2012)
23. Zhang, L., Gao, Y., Hong, C., Feng, Y., Zhu, J., Cai, D.: Feature correlation hypergraph: exploiting high-order potentials for multimodal recognition. IEEE Trans. Cybern. **44**(8), 1408–1419 (2014)

20.

21.

22.

23.

Part IV
Data Analytic Methods and Applications

Analysis Techniques for Feedback-Based Educational Systems for Programming

Nguyen-Thinh Le

Abstract Over the last three decades, many educational systems for programming have been developed to support learning/teaching programming. In order to help students solve programming problems, many educational systems use the means of meaningful feedback that is resulted through an accurate analysis of student programs. In this paper, I review and classify analysis techniques that are required to analyze errors in a student program. This paper also proposes several research directions.

Keywords Educational systems for programming · Feedback · Analysis techniques

1 Introduction

Programming skills are becoming a core competency for almost every profession[1] and thus, Computer Science education is being integrated in the curriculum of almost every study subject. Not only the president of the United States of America[2] and the vice chancellor of Germany, Mr. Gabriel,[3] pleaded in favor of learning coding for everyone, but also the industry also calls for mandatory computer classes at schools. However, it is well-known that programming courses, which constitute an

---

[1]Orsini, L. (2013). Why Programming Is The Core Skill Of The 21st Century? Retrieved 13/01/2015 on: http://readwrite.com/2013/05/31/programming-core-skill-21st-century.

[2]The Hour of Code is here! President Barack Obama wrote his first line of code. Retrieved 13/01/2015 on: http://codeorg.tumblr.com/post/104684466538/hourofcode2014.

[3]Wirtschaftsminister will Programmieren als Fremdsprache. Retrieved 13/01/2015 on http://www.golem.de/news/unterricht-gabriel-will-programmieren-als-fremdsprache-an-schulen-1409-109450.html.

N.-T. Le (✉)
Humboldt-Universität zu Berlin, Berlin, Germany
e-mail: nguyen-thinh.le@hu-berlin.de

© Springer International Publishing Switzerland 2016   141
T.B. Nguyen et al. (eds.), *Advanced Computational Methods for Knowledge Engineering*, Advances in Intelligent Systems and Computing 453, DOI 10.1007/978-3-319-38884-7_11

indispensable part of studies related to Computer Science, are considered a difficult subject by many students [1, 2]. Addressing this problem, various approaches have been proposed to help students learn solving programming problems. One of the solutions to these problems lies with effective technology-enhanced learning and teaching approaches. As a consequence, researchers have identified this gap and have been developing various types of educational systems for programming that are able to analyze students' programming solutions and provide feedback.

Since late 1970s/early 1980s there were the first attempts of developing environments that support students' programming activities. In 1988, Ducassé and Emde [3] reviewed 18 environments and classified them: (1) systems related to tutoring task, (2) general purposes debugging systems, (3) enhanced Prolog tracers. Although the purpose of this review was to classify debugging knowledge and techniques deployed in these systems, there were only seven systems for educational purposes among 18 reviewed systems.

One decade later, Deek and McHugh [4] published a survey in 1998 that reviewed 29 computer-supported systems for learning programming and identified four categories of tools: (1) programming environments, (2) debugging aids, (3) intelligent tutoring systems, and (4) intelligent programming environments. This review identified more types of computer-supported systems for learning programming (programming environments, intelligent tutoring systems, and intelligent programming environments) than in the review of Ducasse and Emde [3]. However, the authors summarized that these systems are limited in two respects: (1) functionality and (2) practical application. That is, in terms of functionality, most of these systems lacked facilities to assist students in performing problem formulation, planning, design and testing of programs. The knowledge base of the intelligent tutoring systems was incomplete. In terms of practical application, the authors reported that there were only few reports on the evaluation of tools or their integration into the classroom.

Since the detailed survey of [4], there were several other smaller surveys on computer-supported systems for learning programming and most of them focused on specific aspects. For instance, Deek et al. [5] reviewed web-based environments for program development. Guzdial [6] investigated programming environments for novices. Douce et al. [7] reviewed computer-supported systems that automatically assess student programming assignments. While these surveys focused on intelligent features (e.g., program analysis, assessment of programming assignments) of systems for supporting students, it is worth noting, on the contrary, Gomez-Albarran [8] focused their survey on less sophisticated and less intelligent approaches. That is, those approaches to learning/teaching programming do not support automatic program analysis nor automatic diagnosis of errors in student programs. Gomez-Albarran reviewed nearly 20 "most outstanding" tools of the following types: (1) tools including a reduced development environment (e.g. BlueJ [9]), (2) example-based environments (e.g. ELM-ART [10]), (3) tools based on visualization and animation (e.g. ANIMAL [11]), and (4) simulation environments (e.g. Karel++ [12]). Another survey that was conducted by Le et al. [13] reviewed

AI-supported instructional approaches for learning programming and classified them into: example-based, simulation-based, collaboration-based, dialogue-based, program analysis-based, and feedback-based approaches.

I see the necessity to investigate analysis techniques that have been deployed in existing educational systems for programming to provide effective feedback due to two reasons. First, feedback is one of the most important features of educational systems (especially, for the domain of programming). It is used by programming learners to revise a solution for a programming problem. Whether a feedback is useful, this depends on feedback's quality (e.g., accurate diagnosis). Second, since the last three decades, not only new systems have been developed, new program analysis approaches have been proposed, while most conducted surveys of educational systems for programming rather focused on classifying the systems [3, 4, 8]. This paper focuses on a wide spectrum of analysis approaches that are used to identify errors in student programs accurately and to provide feedback.

The remainder of this paper is structured as followed. In the next section, I categorize the analysis approaches. In Sect. 3, I discuss the findings of the reviewed analysis approaches. In the last section, I summarize our conclusions and discuss potential research directions.

2   Analysis Techniques for Feedback-Based Educational Systems for Programming

In general, approaches to error diagnosis in programs can be classified into two groups. The first group includes the approaches, which are specific to the domain of programming: plan and bug-based, transformation-based techniques, weighted constraint-based model, and strategy-based model tracing. I use the term "program" to describe the solution for a programming problem when discussing these specific approaches. The second group includes general approaches, which can be applied in general domains: model-tracing, data-mining based, and data-driven approaches.

2.1   Programming-Specific Approaches

2.1.1   Library of Plans and Bugs

Although PROUST [14], ELM-ART [10], and APROPOS2 [15] diagnose errors in programs that are implemented in different programming languages (Pascal, Lisp and Prolog), these systems work on the same principles: (1) modeling the domain knowledge using programming plans and buggy rules, (2) identifying the intention by matching the student program against anticipated programming plans, and (3) detecting errors using buggy rules, which represent common bugs made by students.

In PROUST, each programming problem, which is posed to the student, is represented internally by a set of programming goals and data objects. Programming goals are the requirements, which must be satisfied and data objects are manipulated by the program. A programming plan or an algorithm represents a way to implement a corresponding programming goal. To realize a goal, there might be more than one possible plan.

In contrast to PROUST, APROPOS2 uses the concept of algorithms instead of programming plans. The author argues that the representation of programming plans is suited best for imperative languages like PASCAL where keywords for programming constructs, e.g., FOR-DO, or WHILE-DO, can be used to anchor program analysis. Thus, he proposed to use algorithms as high level concepts in Prolog, and defined this notion as follows: "An algorithm is a particular way of solving a problem that specifies a strategy for the problem's solution but leaves out details of the implementation". According to this definition, an algorithm is comparable to a composition of several programming plans.

Common bugs are normally collected from empirical studies and represented as buggy rules. While PROUST contains only buggy rules, ELM-ART adds two more types: good and sub-optimal rules which are used to comment good programs and less efficient programs (with respect to computing resources or time), respectively. APROPOS2 uses another representation of common bugs which are referred to as buggy clauses. Both buggy clauses and buggy rules serve the same purpose.

In general, a system of this class performs error diagnosis by synthesis in three steps. First, it looks up the problem description and identifies goals to be implemented. Second, it generates a variety of different ways to implement each goal, and derives hypotheses about the plans the programmer may have used to satisfy each goal. Each hypothesis is a possible correct program of the corresponding goal. Third, if the hypothesized plan matches the student program, the goal is implemented correctly. Otherwise, the system looks up the database of buggy rules to explain the plan discrepancies. The procedure of error diagnosis in APROPOS2 exploits algorithms as high level programming concepts instead of programming plans and is carried out in a similar way. First, appropriate algorithms for a given problem are selected, and various possible programs are generated. Second, each generated program is matched against the student program, and the best algorithm is identified using a heuristic search technique. The third step employs buggy clauses to identify errors. In principle, all three systems are able to identify the intention implemented in the student program using plan/algorithm matching, and detect errors using buggy rules/buggy clauses.

This approach has been proven useful to identify the intention underlying student program using programming plans. However, it is often criticized as being laborious, because a programming goal can be implemented according to many different programming plans. In addition, if a programming problem consists of many programming goals to be satisfied, the space of combinations of programming plans would be very large. In addition, specifying buggy rules or buggy clauses requires an extensive study of misconceptions of the students. Such a study normally needs a large corpus of student programs. However, a library of buggy rules or buggy clauses might be specific to a certain population of students. Studies have shown that bug libraries cannot be used effectively with a new student population [16]. Johnson [14] stated that PROUST was able to analyze the intention in 81 % of the programs for the Rainfall problem, which was assigned to a class of novice programmers (maybe at the Yale University, where the author was working). When this system was evaluated at two Belgian universities (UFSIA, Antwerpen and K.U. Leuven Campus Kortrijk), Vanneste [17] reported that this rate dropped to less than 10 %. To improve the ability to analyze programs created by a new student population, PROUST's library of buggy rules needs to be extended considering additional erroneous programming behaviors of the students.

2.1.2   Program Transformation

While the approach of using plan and bug libraries compares student programs to a set of anticipated correct programs and bugs, the transformation approach uses a single reference program to check the correctness of the student's one. The transformation approach can be divided into two classes: program-to-abstraction and program-to-program. In the first class, a student program and a reference program are transformed to higher level abstractions, which are then compared to each other. In the second class, a reference program in normal form is transformed to the best representative one, which is then compared to the student program.

Hong's Prolog [18] belongs to the first class. This system intends to transfer two kinds of programming knowledge to students: high level programming techniques and basic programming concepts of logic programming. The domain model of the system consists of several high level programming techniques and each of them is represented by a set of grammar rules which are used to parse the student program. The system iteratively uses the sets of grammar rules to parse the student program. If the parsing procedure does not finish successfully, that means, the selected set of grammar rules has not been completely exploited and the strategy of the student program cannot be identified. In this case, the system uses one of the possible solution strategies specified for a given problem to guide the student. Otherwise, the solution strategy has been identified, and the system diagnoses errors in the student program. The system uses the same set of grammar rules, with which the solution strategy has been identified, to parse a corresponding reference program. For each possible solution strategy specified for a problem, there is a corresponding reference program. The parse tree of the student program is compared against the one for the reference program. The differences between the two parse trees indicate errors in the student program.

Whereas Hong's Prolog transforms both the student and the reference program to higher-level abstractions, ADAPT transforms a Prolog program to another one [19]. The error diagnosis process is divided into two steps. First, for a given problem, the system begins with a single reference program in normal form, generates a set of representative programs using a set of Prolog schemata. Then the algorithm underlying ADAPT transforms the most appropriate representative program into a

Fig. 1 Transformation-based analysis

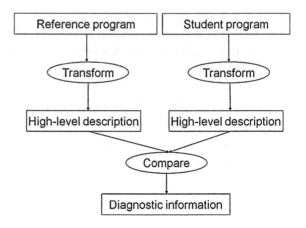

structure that best matches the student program. The second step detects errors by matching the student program and the most representative one. If there are no mismatches, the student program is correct. Otherwise, the system attempts to explain mismatches by searching rules in the bug library. The program-to-program transformation model of ADAPT and the program-to-abstraction transformation model of Hong's tutor are illustrated by Fig. 1.

The transformation approach is very comfortable for the author of an exercise because only one or a few reference programs are required, and specifying a reference program in normal form is a simple task. However, to the best of our knowledge, none of the systems mentioned above has been evaluated with respect to its effectiveness of tutoring. A possible explanation might be that a transformation algorithm is difficult to develop because it has to be verified that the transformed program produces the same results as the initial one. Even if there is a transformation algorithm, it can only be used for a small class of programs. For example, ADAPT is only able to accept a class of programs for the list reversal problem.[4]

2.1.3 Weighted Constraint-Based Model

Instead of the "ideal solution paths" concept used in other tutoring systems, Le and his colleagues [20] proposed to model a semantic table that represent solution strategies for a given problem. Each solution strategy is described by a set of semantic elements required by its implementation. If a semantic element contains an

---

[4]The class of programs is defined by: "(1) remove a single element from the front or back of their input list, (2) use simple variations of standard append 3 for input decomposition and output composition, (3) restrict the use of increasing arguments (i.e., arguments increase in length on each pass) to those that are necessary for the computation (e.g., accumulators) for outputs, (4) use a single recursive clause" [19, p. 8].

arithmetic expression, transformation rules can be used to generate all possible implementation variants. The second modeling technique proposed by Le and colleagues is weighted constraints to compare the semantics between the student program and the requirements of the semantic table and to check general well-formedness conditions. Since for a programming problem there might be many alternative solution strategies, constraints, which are solely based on a binary logic (violated or not), do not contain sufficient information to select the most plausible solution strategy and the most relevant error hypotheses from a multitude of competing ones. Therefore, Le and colleagues enhanced the capability of the classical constraint-based modeling approach [21] by adopting the idea of a probabilistic approach to softening constraints in a constraint satisfaction problem. Constraint-based error diagnosis is a constraint satisfaction problem whose goal is to identify inconsistencies between an erroneous program and a constraint system. Constraint weights are taken from the interval [0, 1], which can be conceived of as a measure of importance with 0 indicating the most important requirements. Weighted constraints allow us to hypothesize the solution strategy possibly pursued by the student, and to examine the semantic correctness of her program. Beyond that, constraint weights can be used to determine the order in which feedback is presented to the student. The authors used semantic tables, weighted constraints, and transformation rules to model knowledge of the domain logic programming. This model has been applied to analyze logic programs.

2.1.4 Strategy-Based Model Tracing

Instead of using a semantic table, which represents alternative solution strategies for a programming problem [20], Gerdes et al. [22] model alternative solution strategies in the system ASK-ELLE through several model programs. Then, the system derives the intended solution strategy of the student using the set of pre-specified model programs. ASK-ELLE uses a strategy language to represent programming strategies [23]. Then, using the identified programming strategy of the model program, many variants are generated. All generated programs are normalized before they are used to compare with the student program. Through the comparison between the student program and the generated programs, the system not only has the ability of analyzing the student's intention underlying a program, but also is able to analyze the student program on the code level. The system is able to provide different types of feedback based on the identified intention of the student. It can provide a description of a particular reference program, propose an implementation for a function, or emphasize one particular implementation method.

2.2   Domain-Independent Analysis Approaches

2.2.1   Model-Tracing

Production rules comprise the core of a model-tracing tutoring system. They represent problem-solving behaviors of experts and typical erroneous behaviors of a population of students. Problem-solving behaviors of experts and typical erroneous behaviors of students are referred to as "ideal" and "buggy" solution paths, respectively [24]. An ideal solution path consists of problem solving steps, which lead to a correct program, while a buggy one anticipates possible erroneous students' problem solving steps. When the student inputs a program, the system monitors her action symbol by symbol and generates a set of possible correct and buggy solution paths. Whenever a student's action can be recognized as belonging to a correct path, the student is allowed to go on. If the student's action deviates from the ideal solution paths, the system generates hints to guide the student towards a correct program. With respect to analyzing errors in student programs, the model-tracing approach has demonstrated the its strengths the educational system JITS for programming Java [25].

2.2.2   Analysis Using Machine Learning

In addition to various analysis techniques that make use of domain knowledge, different machine learning techniques have been applied in order to generate a library of errors that are usually made by novice programmers (JavaBugs [26] and MEDD [27]). Sison et al. [27] applied in MEDD a machine learning technique to learn classes of correct programs, where each class represents a different intention. This technique is based on the exemplar-based approach proposed by Bareiss et al. [28]. The exemplar-based approach determines the similarity between exemplars by matching surface as well as abstract features and exemplars are programs in MEDD. Not only using machine learning to classify correct programs, MEDD also was able to classify incorrect programs in order to represent different classes of knowledge-level errors. For this purpose, an incremental relation clustering technique was used to consider causal relationships in the background knowledge and regularities in the data. That is, the technique clustered discrepancies between incorrect novice programs and their associated intentions (i.e., classes of correct programs) in order to characterize knowledge-level errors. Similarly, JavaBugs [26] used machine learning to build trees of misconceptions using similarity measures and background knowledge.

2.2.3   Data-Driven Approach

Instead of taking much time for modeling expert knowledge (e.g., production rules, ideal solutions, semantic table, properties of solution correctness for a programming problem), the data-driven approach uses a mass of correct student programs. The data-driven approach use correct student programs in order to construct a solution space that contains all solution states students have created in the past (e.g., in the former semesters of a programming course). The solution states build many possible paths to correct solutions.

FIT Java Tutor [29] deploys experts' model programs in addition to correct student programs. The authors proposed to apply machine learning (clustering and classification) techniques in order to select meaningful student programs, which are classified into clusters. By comparing the student's program with one of the pre-specified experts' model programs or solution examples in the clusters of student programs, errors in the student's program can be analyzed. Feedback is generated based on the analysis results. This approach is also applicable in other domains (e.g., sequential data in Chromosomes, Mokbel et al. [30]).

ITAP [31] extended the concept of data-driven to overcome the challenge of modeling open-ended solution space that covers many possible solution paths. For this purpose, the system requires two kinds knowledge: (1) at least one reference program for a given programming problem and (2) a test method that can score, e.g., pairs of expected input and output of code. The process of error analysis and hint generation is based on the combination of algorithms for state abstraction, path construction, and state reification. Based on a reference program, the system constructs a solution space for a problem. The authors defined a solution space as a graph of intermediate states that students pass through as they work on a given problem. Each state is represented by the student's current code. The system starts the process of path construction by inserting new "correct" states into the solution space. The correct states are checked by the test methods. These "correct" states serve to define the optimal path of actions to solve a given problem and to generate feedback by recommending the action a student should take to move from their incorrect state to a correct one. The authors stated that the system is capable to identify errors and to provide hints in any state of student's programs by using a single reference program and the analysis performance of the system (i.e., the accuracy of analysis) can be improved by collecting data over time.

3 Discussion

An interesting finding is that although many various modeling techniques have been devised, but each modeling technique has been used in a single system. That is, the developed modeling techniques do not have widespread use in practice or across several tools for programming education. Several reasons for this phenomenon can be derived. First, each developed modeling technique might be applicable for a specific programming concept or for a specific programming language, whereas the number of programming concepts of a programming paradigm is high and the number of programming language is not predictable. Second, time required for devising a modeling technique for a programming language or a programming concept might be very high, so that researchers, after have tested a devised modeling technique, do not have resources (time or human forces) for testing the devised modeling technique with a new programming language or a new programming concept.

One thing in common between these analysis approaches is that they use at least one reference program to compare the difference between a reference program and a student's program or an abstract representation capturing required components to analyze students' programs (e.g., PROUST, Hong's PROLOG, INCOM). The reference program and the abstract representation of required components are required to understand a student's program and to detect errors in a student's program.

4 Conclusion

In order to be able to provide students appropriate feedback that is intended to scaffold the problem solving process of students during programming, an educational system needs effective one or a combination of several analysis techniques. In this paper, I have reviewed various analysis approaches that have been devised for educational systems for programming. I have identified following major analysis approaches that have been validated and tested. The programming-specific approaches include library of plans and bugs, transformation, weighted constraint-based model, strategy-based model tracing and the domain-independent approaches include model tracing, data mining-based, and data-driven based.

I pointed out that a wide variety of analysis techniques that are required to diagnose student programs and to provide feedback exist. However, mostly they were only used in individual systems. They did not have widespread use in practice, neither were they tested across several tools or domains. As a research direction, researchers could investigate analysis techniques (e.g., model-tracing, weighted constraint-based model, data mining-based) in new domains (i.e., new programming languages or new programming paradigms). This is necessary to prove that a devised analysis technique is universally applicable or limited to a specific domain.

References

1. Altadmri, A., Brown, N.C.: 37 Million compilations: investigating novice programming mistakes in large-scale student data. In: Proceedings of the 46th ACM Technical Symposium on Computer Science Education, pp. 522–527 (2015)
2. Carter, J., Dewan, P., Pichiliani, M.: Towards incremental separation of surmountable and insurmountable programming difficulties. In: Proceedings of the 46th ACM Technical Symposium on Computer Science Education, pp. 241–246 (2015)
3. Ducassé, M., Emde, A.-M.: A review of automated debugging systems: knowledge, strategies and techniques. In: Proceedings of the 10th International Conference on Software Engineering, pp. 162–171. IEEE (1988)

4. Deek, F.P., McHugh, J.A.: A survey and critical analysis of tools for learning programming. Comput. Sci. Educ. **8**(2), 130–178 (1998)
5. Deek, F., Ho, K.-W., Ramadhan, H.: A review of web-based learning systems for programming. In: Montgomerie, C., Viteli, J. (eds.) Proceedings of World Conference on Educational Multimedia, Hypermedia and Telecommunications, pp. 382–387 (2001)
6. Guzdial, M.: Programming environments for novices. In: Fincher, S., Petre, M. (eds.) Computer Science Education Research, pp. 127–154 (2004)
7. Douce, C., Livingstone, D., Orwell, J.: Automatic test-based assessment of programming: a review. J. Educ. Res. Comput **5**(3), 4 (2005)
8. Gomez-Albarran, M.: The teaching and learning of programming: a survey of supporting software tools. Comput. J. **48**(2), 131–144 (2005)
9. Barnes, D.J., Kölling, M.: Objects first with java. Pearson Education (2012)
10. Weber, G., Brusilovsky, P.: ELM-ART—an interactive and intelligent web-based electronic textbook. Int. J. Artif. Intell. Educ. **26**(1), 72–81 (2015)
11. Rösling, G., Freisleben, B.: Animal: A system for supporting multiple roles in algorithm animation. J. Vis. Lang. Comput. **13**(3), 341–354 (2002)
12. Bergin, J., Roberts, J., Pattis, R., Stehlik, M.: Karel++: A Gentle Introduction to the Art of Object-Oriented Programming. Wiley, New York, NY, USA (1996)
13. Le, N.-T., Strickroth, S., Gross, S., Pinkwart, N.: A Review of AI-supported tutoring approaches for learning programming. In: Proceedings of the 1st International Conference on Computer Science, Applied Mathematics and Applications (ICCSAMA), pp. 267–279, Berlin, Germany. Springer, Switzerland (2013)
14. Johnson, W.L.: Understanding and debugging novice programs. Artif. Intell. **42**(1), 51–97 (1990)
15. Looi, C.-K.: Automatic debugging of Prolog programs in a Prolog intelligent tutoring system. Instr. Sci. **20**, 215–263 (1991)
16. Payne, S.J., Squibb, H.R.: Algebra mal-rules and cognitive accounts of error. Cogn. Sci. **14**(3), 445–481 (1990)
17. Vanneste, P.: A reverse engineering approach to novice program analysis. Ph.D. thesis, KU Leuven Campus Kortrijk (1994)
18. Hong, J.: Guided programming and automated error analysis in an intelligent Prolog tutor. Int. J. Hum Comput Stud. **61**(4), 505–534 (2004)
19. Gegg-Harrison, T.S.: Exploiting program schemata in a Prolog tutoring system. Ph.D. thesis, Department of Computer Science, Duke University (1993)
20. Le, N.T., Menzel, W.: Using weighted constraints to diagnose errors in logic programming—the case of an ill-defined domain. Int. J. Artif. Intell. Educ. **19**(4), 382–400 (2009)
21. Ohlsson, S.: Constraint-based student modelling. In: Greer, J.E., McCalla, G.I. (eds.) Student Modelling: The Key to Individualized Knowledge-Based Instruction, pp. 167–189, Springer, Berlin (1994)
22. Gerdes, A., Heeren, B., Jeuring, J., van Binsbergen, L.T.: Ask-Elle: a teacher-adaptable programming tutor for Haskell giving automated feedback. In: Le, N.T., Boyer, K., Hsiao, S. I., Sosnovsky, S., Barnes, T. (eds.) Special Issue on AI-Supported Education in Computer Science. Int. J. Artif. Intell. Educ. (2016)
23. Heeren, B., Jeuring, J., Gerdes, A.: Specifying rewrite strategies for interactive exercises. Math. Comput. Sci. **3**(3), 349–370 (2010)
24. Anderson, J.R., Betts, S., Ferris, J.L., Fincham, J.M.: Neural imaging to track mental states while using an intelligent tutoring system. In: Proceedings of the National Academy of Science, vol. 107, pp. 7018–7023 (2010)
25. Sykes, E.R.: Qualitative evaluation of the java intelligent tutoring system. J. Syst. Cyber (2006)
26. Suarez, M., Sison, R.: Automatic construction of a bug library for object-oriented novice java programmer errors. In: Woolf, B., Aïmeur, E., Nkambou, R., Lajoie, S. (eds.) Intelligent tutoring systems, vol. 5091, pp. 184–193. Springer, Berlin (2008)

27. Sison, R.C., Numao, M., Shimura, M.: Multistrategy discovery and detection of novice programmer errors. Mach. Learn. **38**(1–2), 157–180 (2000)
28. Bareiss, E.R., Porter, B.W., Weir, C.: PROTOS: an exemplar-based learning apprentice (1987)
29. Gross, S., Pinkwart, N.: Towards an integrative learning environment for java programming. In: the IEEE 15th International Conference on Advanced Learning Technologies (ICALT), pp. 24–28 (2015)
30. Mokbel, B., Paaßen, B., Hammer, B.: Adaptive distance measures for sequential data. In: Proceedings of the 22nd European Symposium on Artificial Neural Networks, Computational Intelligence and Machine Learning, pp. 265–270 (2014)
31. Rivers, K., Koedinger, K.R.: Data-driven hint generation in vast solution spaces: a self-improving python programming tutor. In: Le, N.T., Boyer, K., Hsiao, S.I., Sosnovsky, S., Barnes, T. (eds.) Special Issue on AI-Supported Education in Computer Science. Int. J. Artif. Intell. Educ. (2016)

Exploring Drivers of Urban Expansion

Anna Shchiptsova, Richard Hewitt and Elena Rovenskaya

Abstract Spatial patterns in urban land development are linked with the level and type of economic activity. Here, we develop a statistical model to explore the relationship between the spatially explicit population density and the type of land use in a region. The relationship between the type of land use (urban/non-urban) and the level of economic activity is modeled at the scale of a single cell on the geographical map. Thus, the statistical model should be tested against large samples of data points on the high-resolution maps. The challenge here is that the original socio-economic data is given at a coarser resolution than the land use (200×200 m cells) We present results of our spatial modeling exercise for the case study of the Seville Province, Spain.

Keywords Land use model · Urban sprawl · Multiple regression

1 Introduction

In recent decades urban systems have undergone rapid development. We have seen a transition in the population distribution from the population mostly dispersed in rural areas to a highly urbanized society, where people are concentrated in cities. Today more than 50 % of people worldwide live in a city [1] and this figure is likely to grow more in the future.

A. Shchiptsova (✉) · E. Rovenskaya
International Institute for Applied Systems Analysis, Schlossplatz 1,
2361 Laxenburg, Austria
e-mail: shchipts@iiasa.ac.at

R. Hewitt
Observatorio para una Cultura del Territorio, C/ Duque de Fernan Nunez 2,
1a planta, 28012 Madrid, Spain

E. Rovenskaya
Faculty of Computational Mathematics and Cybernetics, Lomonosov Moscow
State University, 2nd Educational Building, Leninskie Gory, 119991 Moscow, Russia

© Springer International Publishing Switzerland 2016   153
T.B. Nguyen et al. (eds.), *Advanced Computational Methods
for Knowledge Engineering*, Advances in Intelligent Systems
and Computing 453, DOI 10.1007/978-3-319-38884-7_12

At the regional scale we observe territory expansion of the urbanized centers. However, this process typically unfolds non-uniformly with respect to the city border. Much of this development has occurred as dispersed, low density growth outside of the major centers but within their area of economic influence. Such type of urban development is typically referred to as urban sprawl [2]. While sprawling developments are not necessarily in themselves always undesirable, they bring a range of issues such as increased energy consumption through encouragement of the use of private vehicles, causing traffic congestion and air pollution, and irreversible damage to ecosystems, caused by scattered and fragmented urban development in open lands [3].

A large body of research is dedicated to the analysis and prediction of urban expansion. Studies on land use change are based on different modeling principles including such techniques as cellular automata [4, 5], Markov chains [6] and logistic regression [7, 8]. This study focuses on the dependence between spatial patterns in land use and population distribution (without the temporal dimension). By using spatial data, we investigate whether part of the variance of the population density is explained by the land use type of the corresponding cell and the types of its immediate neighbors and if, in this way, we can capture spatial interactions.

Geographic data frequently shows spatial dependence, i.e., values at close distances are more similar than expected for independent observations. This property limits the use of the multiple linear regression model for spatial data analysis. An alternative is to incorporate a spatial lag into the model specification (e.g., spatial lag model or spatial error model). A comprehensive introduction to the econometric spatial modeling can be found in [9, 10]. But estimation of the spatial models is not easily computable. This research focuses on application of certain filtering techniques to spatial data in order to meet assumptions of standard linear regression and use conventional statistical methods to test and interpret results of spatial analysis.

We perform a case study on the Spanish Province of Seville. The choice of this region is motivated by the fact that Spain is one of Europe's urban sprawl hotspots, with problems of urban sprawl being particularly acute in the area of economic influence of major cities like Madrid and Valencia and along the Mediterranean coast.

2 Study Area

The Province of Seville is located in the Mediterranean region of Andalusia in the southwestern part of Spain. Its territory is $14,000 \, \text{km}^2$. The terrain in this region is made up almost exclusively of river basin. The Guadalquivir river crosses the province from east to west. In the north territory includes parts of the Sierra Morena mountain range and to the very south the foothills of the Cordillera Subbetica mountain range.

The population of the region is close to two million inhabitants (2010). The province is subdivided into 105 municipalities. The large part of the population lives in the capital city Seville. The Seville municipality has the population of about

700,000 people (2011, INE), which is much larger than any other municipality; for example, the second largest municipality, Dos Hermanas, has the population of about 130,000 inhabitants (2014).

2.1 Land Use Maps

Seville has experienced notable urban development in recent years. In this region, as well as overall in Spain, urban expansion has been especially acute since the restoration of democracy in 1978, joining the EU in 1986 and skyrocketing per capita incomes in the second half of 1980s and the decade before the 2008 crisis; after the 2008 crisis, the speed of development has slowed down.

Figure 1 shows the urban/non-urban land distribution in this region from the year 1956 when Spain was an autocratic country to the post-crisis year 2013. The GIS data represents the territory of the region as a regular grid of cells. We classify all the cells into two major categories: urban land and non-urban land (vegetation, wetlands, agricultural land and water). This figure illustrates the urbanized centers territory expansion unfolding over the last 60 years. Table 1 provides some basic statistics illustrating the spread of urbanization.

2.2 Economic Data

This study focuses on land use distribution in Seville region for the year 2003. As candidate drivers for the land use change in Seville, various socio-economic factors have been identified from the papers analyzing case studies of land conversion in the New Castle County, the USA [8], Wuhan City, China [7], Ecuador [11], San Francisco Bay and Sacramento areas [12] (CUF model), and San Francisco Bay area [4] (SLEUTH model). We have included those drivers of land conversion from these case studies, which are relevant for the Seville province. These potential land use drivers include socio-economic factors defined on the GIS-based maps (with 200×200 m cells) and those obtained from the census data. Where the data for the year 2003 was not available, the closest year, for which the data was available, has been chosen.

The GIS-based maps in the collected dataset include data on transportation networks (i.e., the distance to the nearest road (2005) and the distance to the nearest airport (2006)), biophysical factors (i.e., the distance to the nearest waterfront (2005) and the distance to the nearest area of forest (2006)), data on physical proximity to different infrastructure objects (i.e., the distance to the nearest area of commercial or industrial land use (2006), the proximity to a city center with more than 10,000 inhabitants (2011) and the proximity to a city center with more than 50,000 inhabitants (2011)).

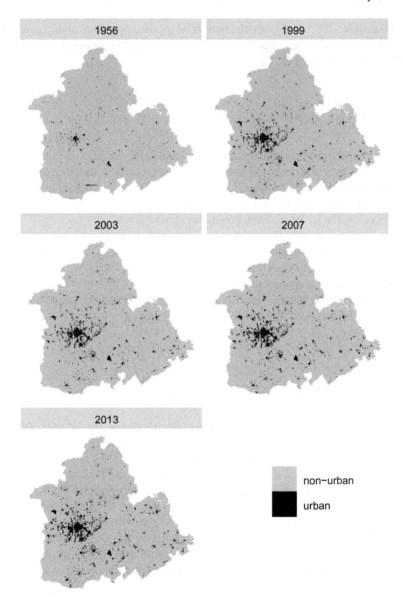

Fig. 1 Historical transformation of land use in the Province of Seville, Spain

Table 1 Spread of urbanization in the Province of Seville estimated from the GIS data

Year	Number of urban cells	% of urban cells in the map (%)	Relative growth (%)
1956	2461	0.71	–
1999	9218	2.63	274.56
2003	10,105	2.88	9.62
2007	12,289	3.50	21.61
2013	13,674	3.89	11.27

The census data on potential factors includes several types of records characterizing the social and economic activity in the region. Namely, the group of general economic factors includes data on the income per capita (2003, Euros, Source: SIMA), people employed (2001, number of people, Source: SIMA), and economically active population (2001, number of people, Source: SIMA). Population factors are composed from the percentage of population younger than 20 years old, the percentage of population between 20 and 64 years old, and the percentage of population older than 65 years (2001, number of people, Source: Instituto de Estadistica y Cartografia de Andalucia, Consejeria de Economia, Innovacion, Ciencia y Empleo). Land economic factors include data on the real estate transactions (2004, number of transactions, Source: Diputacion de Sevilla, Anuario Estadistico de la Provincia de Sevilla) and the number of dwellings built (2001, number of houses, Source: Instituto Nacional de Estadistica). Finally, social factors are represented by the number of secondary schools (2005, number of centers, Source: SIMA).

3 Modeling

Here, we put forward a multiple regression model that relates the expansion of urban territories with the spatial population growth in the following form

$$y = X\alpha + Z_1\beta^1 + Z_2\beta^2 + \gamma + \epsilon$$
$$\epsilon \sim N(0, \sigma^2 I_n) \qquad (1)$$

In (1) y contains a $n \times 1$ vector of the section-based dependent variable, X is a $n \times p$ matrix of the GIS-based explanatory variables describing the land use types of the given cell and its neighboring cells, Z_1 represents a $n \times k_1$ matrix of the socio-economic GIS-based explanatory variables and Z_2 is a $n \times k_2$ matrix of the municipality-based explanatory variables. α, β^1 and β^2 are $p \times 1$, $k_1 \times 1$ and $k_2 \times 1$ vectors of the corresponding coefficients respectively and γ is an intercept. The error term ϵ is a $n \times 1$ vector of independent identically normally distributed variables with mean equal zero and variance equal σ^2. I_n denotes a $n \times n$ identity matrix. Observations are accounted at the level of a single cell. Thus, the number of cells defines the sample size n.

We take the population density as dependent variable in the regression equation (1). Note that population density is defined at the lowest level of administrative division for the census data in Spain (census tracts called sections). As the section level is coarser than the cell level, we assign the value of the population density in a given section to every cell that belongs to this particular section. In the same way, we also extend the municipality-level socio-economic data to the level of a single cell on a GIS-based map.

Note also, that in general, X includes the type of land use in a focal cell and the types of land use in its Moore neighborhood, which comprises the eight cells surrounding a central cell on a two-dimensional square lattice. Alternatively, the information about the cell neighborhood can be aggregated and represented just by a number of urban cells in it (including the type of a focal cell). In what follows, we employ the latter aggregation for defining X.

3.1 Implementation

The computations are done in the Clojure programming language (version 1.6.0). Regression estimates are obtained using Incanter 1.9.0, a Clojure-based, R-like statistical computing and graphics environment for the JVM. For principal component analysis, we use R version 3.0.3.

The input GIS-based maps are stored in an ESRI ArcInfo ASCII raster file format. The census data is given as tabular records in a csv file format.

3.2 Pre-processing of the GIS-based Explanatory Variables

Before performing the regression analysis, we rescale and clean the source GIS-based maps. We use the rule that if either any of neighboring land use types of a cell are undefined or socio-economic data is not set for the corresponding section or municipality, we exclude the cell from the sample. The algorithm of GIS data cleaning is done sequentially:

Step 1. Exclusion of cells with neighborhoods containing undefined values—we exclude cells, which either belong to the border of the studied area or have undefined values in their neighborhood in any of the given maps in the dataset.

Step 2. Normalization of the GIS-based explanatory variables—we bring all the values of these variables into the range [0, 1].

Step 3. Exclusion of cells, which fall into protected natural areas in the Seville province, or cells whose neighborhood contains cells belonging to these areas.

Step 4. Exclusion of cells, which do not have a specified value of the population density in the section they belong to.

Note that protected natural areas include UNESCO World Heritage sites, Ramsar wetland sites, Nature network 2000 sites, biosphere reserves, protected areas and European Diploma sites. Since urban development is not allowed at all of these sites, they have been excluded from the sample.

3.3 Principal Component Analysis

First, we find out that the municipality-based variables exhibit strong pairwise correlation as shown in Table 2. Note that in case of moderate- and big-size samples (i.e., those containing more than 80 points), the critical Pearson correlation coefficient that ensures the statistical significance at 0.05 level is near 0.25.

Because of a large number of highly correlated variables, the principal component analysis is applied to reveal interrelationships and remove multicollinearity in the set of the municipality-based economic factors. This kind of transformation of the original variables allows obtaining orthogonal factors, which certainly do not correlate with each other.

In this case, we are able to reduce the number of factors to the first two principal components, which explain 99 % of the total sample variance. The first component represents the average yield of 9 out of 10 variables. The second principal component correlates with the remaining original explanatory variable—income per capita. The intuition behind the revealed two first principal components is the following. The first principal component separates municipalities with a high number of inhabitants (by assigning bigger values) from the underpopulated territories. The second principal component separates observations from the municipalities with medium population (in terms of the Seville Province) from other points in the sample.

3.4 Data Compilation

After the procedure described in Sect. 3.2, we have a dataset with 235,678 observations, which fall into 99 different municipalities with 1,219 sections. In contrast to the GIS data, municipalities and sections do not divide territory uniformly and vary in size significantly. In highly populated areas section size coincides with one cell, while the upper value of the size can exceed 20,000 cells in other territories.

Table 3 contains a list of variables, which serve as an input to the multiple regression analysis.

Table 2 Correlations of the municipality-based variables ($n = 99$)

	Income per capita	Pop employed	Pop active	Pop under 20	Pop from 20–65	Pop over 65	Real estate	Dwellings built	Schools
Income per capita	1	−0.06	−0.06	−0.07	−0.06	−0.05	−0.06	−0.12	−0.09
Pop employed	−0.06	1	1	1	1	1	0.99	0.96	0.99
Pop active	−0.06	1	1	1	1	1	0.99	0.96	0.99
Pop under 20	−0.07	1	1	1	1	0.99	0.99	0.97	0.99
Pop from 20–65	−0.06	1	1	1	1	1	0.99	0.96	0.99
Pop over 65	−0.05	1	1	0.99	1	1	0.99	0.94	0.99
Real estate	−0.06	0.99	0.99	0.99	0.99	0.99	1	0.96	0.99
Dwellings built	−0.12	0.96	0.96	0.97	0.96	0.94	0.96	1	0.97
Schools	−0.09	0.99	0.99	0.99	0.99	0.99	0.99	0.97	1

Table 3 Variables included in the multiple regression analysis

Variable	Range
GIS-based	
Number of cells with urban land use in the neighborhood	$0 \ldots 9$
Distance to the nearest road	$[0, 1]$
Distance to the nearest area of commercial or industrial land use	$[0, 1]$
Distance to the nearest airport	$[0, 1]$
Distance to the nearest waterfront	$[0, 1]$
Distance to the nearest area of forest	$[0, 1]$
Distance to the nearest city with more than 10,000 inhabitants	$[0, 1]$
Distance to the nearest city with more than 50,000 inhabitants	$[0, 1]$
Municipality-based	
PC1 (first principal component)	$[-28.44, 0.89]$
PC2 (second principal component)	$[-4, 1.45]$
Section-based	
Population density (people per cell, 2001)	$[0, 2704]$

4 Results

In what follows, we take the log transformed population density as the dependent variable in the model.

The advantage of the classical multiple linear regression is that we can easily interpret the estimated coefficients if the standard assumptions of the model are fulfilled. However, the latter is rarely a case for the spatial data, and this study is not an exception. Figure 2c shows that residuals are not statistically independent and substantial heteroscedasticity is present when we apply regression (1) to all cells from the GIS-based maps which remain in the dataset after the pre-processing step described in Sect. 3.2.

A non-uniform administrative division of the territory is one of the reasons that may cause the violation of the error independence assumption. Sections necessarily represent the entire province, including areas where human economic activity is very low due to difficult or undeveloped terrain. As a rule, these sections contain many cells (more than 100) and are characterized by a low population density and monotonic change in the distance-related explanatory variables.

Unlike in, e.g., [9, 11], we do not use more advanced statistical models that can incorporate spatial autocorrelation in these areas, but apply a certain filtering to deal with non-independence and heteroscedasticity of the residuals. For this purpose, two filters have been constructed:

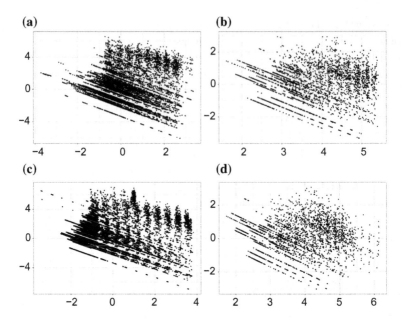

Fig. 2 Residuals (y-axis) versus predicted values (x-axis). **a** Filter 1—Model 1. **b** Filter 2—Model 1. **c** All cells—Model 1. **d** Filter 2—Model 2

Filter 1.   Exclude all cells, which have either all non-urban or all urban cells in their Moore neighborhoods (including the focal cell itself).

Filter 2.   Exclude all cells, which have all urban neighbors around them (including the focal cell itself) as in Filter 1, and additionally, exclude all sections, where the share of non-urban cells in the total number of cells is more than 70 % and the number of non-urban cells in this section exceeds 100.

Note that in both filters we exclude the urbanized cells with entirely urbanized neighborhoods, because the population density in these cells is likely to be highly dependent on some other (missing in this study) independent variables (for all urbanized cells with entirely urbanized neighborhoods, the distance to roads remains zero and the distance to commercial centers changes insignificantly to capture the population density variance).

4.1   Results of Multiple Regression

In the current exploratory research, we perform the regression analysis applying either Filter 1 or Filter 2, and also varying the number of the GIS-based explanatory variables. Figure 2a, b shows the residual plots against predicted values after applying Filter 1 and Filter 2 correspondingly. In case of the original GIS-based maps,

Table 4 Multiple regression results (Filter 2)

Variable	Coefficient
Model 1	
Number of urban neighbors	0.23 (0.22, 0.24)
Distance to roads	−8.42 (−10.86, −5.97)
Distance to the commercial centers	−1.46 (−2.44, −0.49)
PC1	0.02 (0.01, 0.02)
PC2	0.29 (0.26, 0.32)
R^2 (adjusted R^2)	0.42 (0.42)
Model 2	
Number of urban neighbors	0.22 (0.21, 0.23)
Distance to roads	−6.65 (−9.06, −4.25)
Distance to the commercial centers*	−0.07 (−1.02, 0.88)
Distance to airport	−1.45 (−2.28, −0.61)
Distance to waterfront	−0.35 (−0.7, 0.0)
Distance to forest	0.52 (0.31, 0.73)
Distance to the city with >10 ths people	−1.59 (−2.06, −1.13)
Distance to the city with >50 ths people	3.05 (2.27, 3.82)
PC1	0.01 (0.0, 0.02)
PC2	0.28 (0.25, 0.31)
R^2 (adjusted R^2)	0.47 (0.46)

*Not significant (p-value > 0.05)

the sample size n is 235,678 cells (case c in Fig. 2). Filter 1 reduces this number to 17,351 cells, while Filter 2 leaves 4,138 points for the regression analysis.

Here, we present two models, which differ in the number of explanatory variables. Model 1 includes two GIS-based explanatory variables, which have a visually identifiable correlation trend with the population density on the respective scatterplots. Model 2 takes all available factors as explanatory variables. Figure 2d presents the residual plot for Model 2 after Filter 2 has been applied.

Table 4 presents the estimates obtained from Model 1 and Model 2. In both cases we use the sample obtained from Filter 2.

4.2 Discussion

Figure 2 illustrates that filtering of the original observations facilitates the fulfillment of the standard multiple regression assumptions regarding the error independence and homoscedasticity. Note that a well-recognized problem in the analysis of spatial data is the presence of spatial autocorrelation in the dataset. The effects of spatial dependence on the conventional statistical methods are various, including a likely overestimation of R^2 and the unreliability of the significance tests. Filtering helps to deal with the spatial autocorrelation in such a way, that we select a sample of (presumably) independent observations (cells) from the original GIS-based land use map consequently removing spatial correlations in the sample.

Coefficient estimates in Table 4 indicate that the population density is higher in the cells with more urbanized neighborhoods and is lower in the cells which are far away from the transportation routes. The closer a cell is to commercial centers, the higher the population density is, but this variable has a lesser impact. All estimates in Table 4 are statistically significant (p-value < 0.05).

The R^2 coefficient suggests that about 40 % of the total variance in the population density is explained by Model 1 using Filter 2. This value almost doubles compared to the case when Filter 1 is used. Despite the fact that we cannot explain most of the variance of the population density using the collected set of socio-economic indicators inside the urban areas (and for the whole territory in general), the results obtained so far suggest, that the land use neighborhood partly captures the spatial pattern of the population distribution, which is caused by unknown drivers not included in this study.

5 Conclusion

We have showed that filtering procedures can be used to deal with a non-independence of observations in case of spatial data. Heteroscedasticity and correlation have been detected in the residuals of the standard multiple regression model in the case study of the Seville province. We have constructed special filters to isolate the territory at the borders of urban areas, where spatial correlation is not present. The obtained results have showed a more random pattern in the residuals of the regression model.

The model suggests that for the remaining part of the spatial data (after filtering), a decent part of the variance of the population density can be explained by the abundant land use types in the neighborhood of the focal cell. This finding can help obtain new insights related to the phenomenon of urban sprawl, which occurs at the fringe of urban areas outside of the city centers with high population densification.

Acknowledgments The authors would like to acknowledge DG research for funding through the FP7-funded COMPLEX project #308601, http://www.complex.ac.uk.

References

1. United Nations Population Division: World Population Prospects: The 2004 Revision and World Urbanization Prospects: The 2005 Revision. United Nations, New York (2005)
2. Angel, S., Sheppard, S.C., Civco, D.L.: The Dynamics of Global Urban Expansion. Transport and Urban Development Department, The World Bank, Washington, DC (2005)

3. Frenkel, A., Ashkenazi, M.: Measuring urban sprawl: how can we deal with it? Environ. Plann. B Plann. Des. **35**(1), 56–79 (2008)
4. Clarke, K.C., Hoppen, S., Gaydos, L.J.: A self-modifying cellular automaton model of historical urbanization in the San Francisco Bay Area. Environ. Plann. B Plann. Des. **24**, 247–261 (1997)
5. Batty, M., Xie, Y., Sun, Z.: Modeling urban dynamics through GIS-based cellular automata. Comput. Environ. Urban Syst. **23**, 205–233 (1999)
6. Lopez, E., Bocco, G., Mendoza, M., Duhau, E.: Predicting land-cover and land-use change in the urban fringe: a case in Morelia City Mexico. Landscape Urban Plann. **55**, 271–285 (2001)
7. Cheng, J., Masser, I.: Urban growth pattern modeling: a case study of Wuhan City PR China. Landscape Urban Plann. **6**, 199–217 (2003)
8. Huang, B., Zhang, L., Wu, B.: Spatiotemporal analysis of rural-urban land conversion. Int. J. Geograph. Inf. Sci. **23**(3), 379–398 (2009)
9. Anselin, L.: Spatial Econometrics: Methods and Models. Kluwer Academic Publishers, Dordrecht (1988)
10. LeSage, J., Pace,R.K.: Introduction to Spatial Econometrics. CRC Press (2009)
11. Overmars, K.P., de Koning, G.H.J., Veldkamp, A.: Spatial autocorrelation in multi-scale land use models. Ecol. Modell. **164**, 257–270 (2003)
12. Landis, G.D.: Imagining land use futures: applying the California urban futures model. J. Am. Plann. Assoc. **61**(4), 438–457 (1995)

The Cropland Capture Game: Good Annotators Versus Vote Aggregation Methods

Artem Baklanov, Steffen Fritz, Michael Khachay, Oleg Nurmukhametov and Linda See

Abstract The Cropland Capture game, which is a recently developed Geo-Wiki game, aims to map cultivated lands using around 17,000 satellite images from the Earth's surface. Using a perceptual hash and blur detection algorithm, we improve the quality of the Cropland Capture game's dataset. We then benchmark state-of-the-art algorithms for an aggregation of votes using results of well-known machine learning algorithms as a baseline. We demonstrate that volunteer-image assignment is highly irregular and only good annotators are presented (there are no spammers and malicious voters). We conjecture that the last fact is the main reason for surprisingly similar accuracy levels across all examined algorithms. Finally, we increase the estimated consistency with expert opinion from 77 to 91 % and up to 96 % if we restrict our attention to images with more than 9 votes.

Keywords Crowdsourcing · Image processing · Votes aggregation

A. Baklanov (✉) · S. Fritz · L. See
International Institute for Applied Systems Analysis (IIASA), Laxenburg, Austria
e-mail: baklanov@iiasa.ac.at

S. Fritz
e-mail: fritz@iiasa.ac.at

L. See
e-mail: see@iiasa.ac.at

A. Baklanov · M. Khachay · O. Nurmukhametov
Krasovsky Institute of Mathematics and Mechanics, Ekaterinburg, Russia
e-mail: mkhachay@imm.uran.ru

O. Nurmukhametov
e-mail: oleg.nurmuhametov@gmail.com

A. Baklanov · M. Khachay
Ural Federal University, Ekaterinburg, Russia

© Springer International Publishing Switzerland 2016
T.B. Nguyen et al. (eds.), *Advanced Computational Methods for Knowledge Engineering*, Advances in Intelligent Systems and Computing 453, DOI 10.1007/978-3-319-38884-7_13

167

1 Introduction

Crowdsourcing is a new approach for solving data processing problems for which conventional methods appear to be inaccurate, expensive, or time-consuming. Nowadays, the development of new crowdsourcing techniques is mostly motivated by so called Big Data problems, including problems of assessment and clustering of large datasets obtained in aerospace imaging, remote sensing, and even in social network analysis. For example, by involving volunteers from all over the world, the Geo-Wiki project tackles the problems of environmental monitoring with applications to flood resilience, biomass data analysis and forecasting, etc. The Cropland Capture game, which is a recently developed Geo-Wiki game, aims to map cultivated lands using around 17,000 satellite images from the Earth's surface. Despite recent progress in image analysis, the solution to these problems is hard to automate since human-experts still outperform the majority of learnable machines and other artificial systems in this field. Replacement of rare and expensive experts by a team of distributed volunteers seems to be promising, but this approach leads to challenging questions: how can we aggregate individual opinions optimally, obtain confidence bounds, and deal with the unreliability of volunteers?

The main goals of the Geo-Wiki project are collecting land cover data and creating hybrid maps [14]. For example, users answer 'Yes' or 'No' to the question: 'Is there any cropland in the red box?' in order to validate the presence or absence of cropland [13]. In the paper [1], which is related to use of Geo-Wiki data, researchers studied the problem of using crowdsourcing instead of experts. The research showed that it is possible to use crowdsourcing as a tool for collecting data, but it is necessary to investigate issues such as how to estimate reliability and confidence.

This paper presents a case study that aims to compare the performance of several state-of-the-art vote aggregation techniques specifically developed for the analysis of crowdsourcing campaigns using the image dataset obtained from the Cropland Capture game. As a baseline, some classic machine learning algorithms such as Random Forest, AdaBoost, etc., augmented with preliminary feature selection and a preprocessing stage, are used.

The rest of the paper is structured as follows. In Sect. 2, we give a brief overview of efforts related to the vote aggregation in crowdsourcing. In Sect. 3, we describe the general structure of the dataset under consideration. In Sect. 4, we propose quality improvements for the initial image dataset, introduce our vote aggregation heuristic and existing state of the art algorithms. Finally, in Sect. 5, we analyse the dataset and present our benchmarking results. Then we use annotator models to classify volunteers.

2 Related Work

In the theoretical justification of crowdsourcing image-assessment campaigns, there are two main problems of interest. The first one is the problem of ground truth estimation from crowd opinion. The second one, which is equally important, deals with the individual performance assessment of the volunteers who participated in the campaign. The solution to this problem is in the clustering of voters with respect to their behavioural strategies into groups of *honest workers*, *biased annotators*, *spammers*, *malicious users*, etc. Reflection of this posterior knowledge by reweighting of individual opinions of the voters can substantially improve the overall performance of the aggregated decision rule.

There are two basic settings of the latter problem. In the first setup, a crowdsourcing campaign admits some quantity of images previously labeled by experts (these labels are called *golden standard*). In this case, the problem can be considered as a supervised learning problem, and for its solution, conventional algorithms of ensemble learning (for example, boosting [6, 10, 18]) can be used. On the other hand, in most cases, researchers deal with the full (or almost full) absence of labeled images; ground truth should be retrieved simultaneously with estimation of voters' reliability, and some kind of unsupervised learning techniques should be developed to solve the problem.

Prior works in this field can be broadly classified in two categories: EM-algorithm inspired and graph-theory based. The works of the first kind extend results of the seminal paper [2], applying a variant of the well known EM-algorithm [3] to a crowdsourcing-like setting of the computer-aided diagnosis problem. For instance, in [12], the EM-based framework is provided for several types of unsupervised crowdsourcing settings (for categorical, ordinal and even real answers) taking into account different competency level of voters and different levels of difficulty in the assessment tasks. In [11], by proposing a special type of prior, this approach is extended to the case when most voters are *spammers*. Papers [7, 9, 15] develop the fully unsupervised framework based on Independent Bayesian Combination of Classifiers (IBCC), Chinese Restaurant Process (CRP) prior, and Gibbs sampling. Although EM-based techniques perform well in many cases, usually, they are criticized for their heuristic nature since in general there are no guarantees that the algorithm finds a global optimum.

Furthermore, in [5], an efficient reputation algorithm for identifying adversarial workers in crowdsourcing campaigns is elaborated. For some conditions, the reputation scores proposed are proportional to the reliabilities of the voters given that their number tends to infinity. Unlike the majority of EM-based techniques, the listed results have solid theoretical support, but conditions for which their optimality is proven (especially the graph-regularity condition) are too restrictive to apply them straightforward in our setup.

A highly intuitive and computationally simple algorithm, Iterative Weighted Majority Voting (IWMV), is proposed in [8]. Remarkably, for its one step version, theoretical bounds on the error rate is obtained. Experiments on synthetic and real-life data (see [8]) demonstrate that IWMV performs on a par with the state-of-the-art algorithms and around one hundred times faster than the competitors. Since the dataset of the Cropland Capture Game contains around 4.6 million votes, this computational effectiveness makes IWMV the most suitable representative among the state-of-the-art methods for the benchmark.

The aforementioned arguments have motivated us to carry out a case study on the applicability of several state-of-the-art vote aggregation techniques to an actual dataset obtained from the Cropland Capture game. Precisely, we compare the classic EM algorithm [2], methods proposed in [5], IWMV [8], and a heuristic based on the computed reliability of voters. As a baseline, we use the simple Majority Voting (MV) heuristic and several of the most popular universal machine learning techniques.

3 Data

The results of the game were captured as shown in two tables. The first table contains details of the images: *imgID* is an image identifier; *link* is the URL of an image; *latitude* and *longitude* are geo-coordinates which refer to the centroid of the image; *zoom* is the resolution of an image (values: 300, 500, 1000 m). The following table shows some sample of image data.

imgID	link	latitude	longitude	zoom
3009	http://cg.tuwien.ac.at/~sturn/crop/img_-112.313_42.8792_1000.jpg	42.8792	−112.313	1000
3010	http://cg.tuwien.ac.at/~sturn/crop/img_-112.313_42.8792_500.jpg	42.8792	−112.313	500
3011	http://cg.tuwien.ac.at/~sturn/crop/img_-112.313_42.8792_300.jpg	42.8792	−112.313	300

All votes, i.e. 'a single decision by a single volunteer about a single image' [13], were collected in the second table: *ratingID* is a rating identifier; *imgID* is an image identifier; *volunteerID* is a volunteer's identifier; *timestamp* is the time when a vote was given; *rating* is a volunteer's answer. The possible values for *rating* are as follows: 0 ('Maybe'), 1 ('Yes'), −1 ('No'). The following table shows some sample of vote data.

ratingID	imgID	volunteerID	timestamp	rating
75811	3009	178	2013-11-18 12:50:31	1
566299	3009	689	2013-12-03 08:10:38	0
641369	3009	1398	2013-12-03 17:10:39	−1
3980868	3009	1365	2014-04-10 16:52:07	1

During the crowdsourcing campaign, around 4.6 million votes were collected. We convert the votes to a rating matrix. The matrix consists of ratings given to images (matrix columns) by the volunteers (matrix rows)

$$R = (r_{v,i})_{v=1,i=1}^{|V|,|I|},$$ (1)

V—the set of all volunteers; I—the set of all images with at least 1 vote; $r_{v,i}$—a vote given by a volunteer to an image. Due to an unclear definition, the *'Maybe'* answer is hard to interpret. As a result, we treat *'Maybe'* as a situation when the user has not seen the image; both situations are coded as 0. If a volunteer has multiple votes for the same image, then *only the last vote is used*.

4 Methodology

4.1 Detection of Duplicates and Blurry Images

Since the dataset collected via the game was formed by combining different sources, it is possible that almost the same images can be referenced by different records. In order to check this, we download all 170041 .jpeg images (512*512 size). The total size of all images is around 9 GB. Then we employ perceptive hash functions to reveal such cases. Examples of such functions are aHash (Average Hash or Mean Hash), dHash, and pHash [17]. Perceptual hashing aims to detect images such that a human cannot see the difference. We find that pHash performs much better than computationally less expensive dHash and aHash methods. Note that for a fixed image, the set of all images that is similar according to pHash will contain all images with the corresponding MD5 or SHA1 hash. To summarize, we detect duplicates for 8300 original images; votes for duplicates were merged.

Accepting the idea of the wisdom of the crowd, in order to make a better decision for an image, we need to collect more votes for each image. The detection of all similar images increases statistically significant effects and decreases the dimensionality of the data. In addition, if the detection is performed before the start of the campaign, there is a reduction in the workload of the volunteers.

A visual inspection of images shows the presence of illegible and blurry (unfocused) images. As expected, these images bewildered the volunteers. Thus, we apply automatic methods for blur detection. Namely, by using the Blur Detection algorithm [16], we detect 2300 poor quality images such that it is not possible to give the right answers even for experts. Note that for those images, voting inconsistency is high; volunteers change their opinions frequently. After consultation with the experts, we remove all images of poor quality to decrease the noise level and uncertainty in the data. Finally, we reduce number of images from 170041 to 161752. Note that $|I| = 161752$ and $|V| = 2783$.

4.2   Majority Voting Based on Reliability

In this subsection we present a conjunction of MV and the widely used notion of reliability (see, for example [5]). It is a standard to define reliability w_v of volunteer v as

$$w_v = 2p_v - 1$$

where p_v is the probability that v gives a correct answer; it is assumed that it does not depend on the particular task. Obviously, $w_v \in [-1, 1]$. We use traditional weighted MV with weights obtained by the above rule. The heuristic admits a refinement; one may iteratively remove a volunteer with the highest penalty, then recalculate penalties, and obtain new results for the weighted MV.

Algorithm 1 Weighted MV

Input: V — the set of all volunteers;
I — the set of all images with at least 1 vote;
$R = \left(r_{v,i}\right)_{v=1,i=1}^{|V|,|I|}$ — rating matrix (see (1));
E — the set of images with ground truth labels;
$(e_i)_{i \in E} \in \{-1; 1\}^{|E|}$ — ground truth labels for images from E.
Initialization:
    for $v \in V$: **do**
        if $\sum_{i \in I \cap E} \mathbf{1}(r_{v,i} \neq 0) \neq 0$ **then**
            $w_v \leftarrow 2 \times \frac{\sum_{i \in I \cap E} \mathbf{1}(r_{v,i} = e_i)}{\sum_{i \in I \cap E} \mathbf{1}(r_{v,i} \neq 0)} - 1$
        else
            $w_v \leftarrow 0$
Repeat
Calculate penalties for volunteers according to Algorithm 2. The algorithm takes I, V, R as inputs and gives a vector $(d_v)_{v \in V}$ as output. For volunteer $\hat{v}$ with the highest penalty, we set

$$w_{\hat{v}} \leftarrow 0,$$

$$r_{\hat{v},i} \leftarrow 0 \; \forall i \in I$$

Until convergence or reaching S iterations
Output: the predicted labels $(y_i)_{i \in I}$

$$y_i = \arg\max_{k \in \{-1;1\}} \sum_{v \in V} w_v \mathbf{1}(r_{v,i} = k) \; \forall i \in I$$

The proposed heuristic is presented in Algorithm 1 relying on Algorithm 2 [5, Hard penalty], for which we also provide pseudocode. We now briefly describe an optimal semi-matching (OSM) used in Algorithm 2. Let $B = (N_{left} \cup N_{right}, E)$ be a bipartite graph with N_{left} the set of left-hand vertices, N_{right} the set of right-hand vertices, and edge set $E \subset N_{left} \times N_{right}$. A *semi-matching* in B is a set of edges $M \subset E$ such that each vertex in N_{right} is incident to exactly one edge in M. We stress that it is possible for vertices in N_{left} to be incident to more than one edge in M.

Algorithm 2 Hard penalty [5]

Input: V — the set of all volunteers;

I — the set of all images with at least 1 vote;

$R = \left(r_{v,i}\right)_{v=1,i=1}^{|V|,|I|}$ — rating matrix (see (1));

Initialization:

Obtain I^{cs}, the set of all (conflict) images with both labels (1 and −1);

Create a bipartite graph B^{cs} as follows:

 1: Each volunteer $v_i \in V$ is represented by a node on the left;
 2: Each image $t_j \in I^{cs}$ is represented by two nodes on the right t_j^+ and t_j^-
 3: For each volunteer v_i and image t_j:

    if $r_{i,j} = 1$ **then**
      add the edge (v_i, t_j^+)
    else if $r_{i,j} = -1$ **then**
      add the edge (v_i, t_j^-).

Compute an optimal semi-matching (OSM) [4] for B^{cs} :

    for $v_i \in V$: **do**
      $d_i \leftarrow$ degree of volunteer v_i in the OSM

Output: Penalties for volunteers $\{d_1, d_2, \ldots, d_{|V|}\}$.

An OSM is defined using the following cost function. For $u \in N_{left}$, let $deg_M(u)$ denote the number of edges in M that are incident to u and let

$$cost_M(u) = \sum_{i=1}^{deg_M(u)} i = \frac{(deg_M(u) + 1)deg_M(u)}{2}.$$

An *optimal semi-matching* then, is one which minimizes $\sum_{u \in N_{left}} cost_M(u)$. This definition is inspired by the load balancing problem studied in [4]. We also benchmark the Iterative Weighted Majority Voting algorithm (IWMV) introduced in [8]; see pseudocode of Algorithm 3.

5 Benchmark

To evaluate the volunteers' performance, a part of the dataset (854 images) was annotated by an expert after the campaign took place. For these images 1813 volunteers gave 16,940 votes in total. We then sampled two subsets for training and testing (70/30 ratio). We now empirically assess the irregularity of volunteer-image assignment in the expert dataset. First, using Fig. 1, we answer the question '*How many volunteers voted some specified number of times?*' Second, using Fig. 2, we answer the question '*What is the percentage of volunteers who voted some specified number*

Algorithm 3 The iterative weighted majority voting (IWMV) [8]

Input: V — the set of all volunteers;
I — the set of all images with at least 1 vote;
$R = \left(r_{v,i}\right)_{v=1,i=1}^{|V|,|I|}$ — rating matrix (see (1));
Initialization: $w_v = 1 \; \forall v \in V$; $T_{v,i} = \mathbf{1}(r_{v,i} \neq 0) \; \forall v \in V, \; \forall i \in I$;
Repeat

$$y_i \leftarrow \underset{k \in \{-1;1\}}{\arg\max} \sum_{v \in V} w_v \mathbf{1}(r_{v,i} = k) \; \forall i \in I, \tag{2}$$

$$p_v \leftarrow \frac{\sum_{i \in I} \mathbf{1}(r_{v,i} = y_i)}{\sum_{i \in I} T_{v,i}}, \quad w_v \leftarrow 2p_v - 1 \; \forall v \in V.$$

Until convergence or reaching S iterations;
Output: the predicted labels $(y_i)_{i \in I}$ (see (2)).

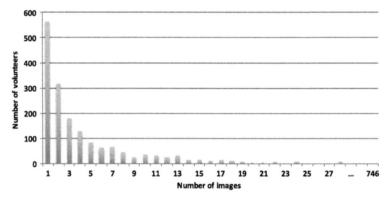

Fig. 1 The figure contains information on the images per volunteer distribution and answers the question '*How many volunteers voted some specified number of times?*'. Namely, the number of images assessed by a volunteer is on the horizontal axes. The number of volunteers is on the vertical axes. Thus, one can see that 566 volunteers labeled only one image from the expert dataset

of times or less?'. Finally, by means of Fig. 3, we answer the questions '*How many images were labeled a specified number of times?*' and '*What percentage of images were labeled some specified number of times or less?*'.

The baseline. To use some conventional machine learning algorithms, *we first apply SVD to the whole dataset* to reduce dimensionality. A study of the explained variance helps us to make an appropriate choice for the number of features: 5, 14, 35. Then we transform the feature space of the testing and training subsets accordingly. On the basis of 10-fold cross-validation of the training subset, we fit parameters for the AdaBoost and Random Forest algorithms. For Linear Discriminant Analysis (LDA), we use default parameters. The accuracy of the algorithms with fitted parameters was estimated using the testing subset; see Table 1.

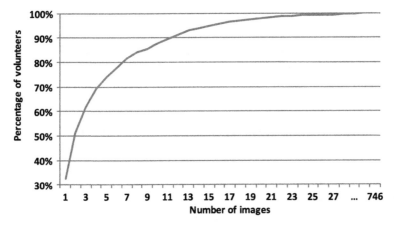

Fig. 2 The figure contains information on the images per volunteer distribution and answers the question '*What is the percentage of volunteers who voted some specified number of times or less?*'. Namely, the maximum number of images assessed by a volunteer is on the horizontal axes. The number of volunteers is on the vertical axes. Thus, one can see that around 49 % of all volunteers saw 2 images or less from the experts dataset. Moreover, around 70 % of volunteers saw 5 images or less from the experts dataset; 90 % of volunteers saw 15 images or less; and 95 % of volunteers saw less than 26 images

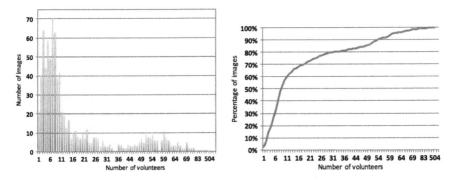

Fig. 3 The figure contains information on volunteers per image distribution. The left plot answers the question '*How many images were labeled a specified number of times?*'. Here one can find that 71 images were labeled by exactly 7 volunteers, 11 images were labeled by 59 volunteers, and 5 images were labeled by 69 volunteers. The right plot answers the question and '*What percentage of images were labeled some specified number of times or less?*'. One can see that around 27 % of images have 5 votes or less, 52 % of images have 9 votes or less, and 75 % of images have 25 votes or less

Benchmarking of algorithms for an vote aggregation is performed as follows. We feed the expert dataset to the algorithms and check their accuracy on the same test subset as above. Note that the transformation of a feature space is not required in this case. In this section, we experimentally test the heuristic based on reliability and compare it with the state-of-art algorithms designed for crowdsourcing. We

Table 1 Accuracy of baseline algorithms corresponding to different number of features

Number of features	Random forest	LDA	AdaBoost
5	**89.92**	87.60	89.15
14	89.14	**90.70**	89.92
35	88.37	89.53	**91.08**

Table 2 Accuracy for 'crowdsourcing' algorithms without image thresholding

	MV	EM	Weighted MV	IWMV
Base	89.81	89.81	90.63	89.81
1	90.05	90.16	**91.45**	89.81
2	90.05	90.05	**91.45**	–
3	89.67	89.58	91.22	–
4	89.34	89.46	90.98	–
5	89.93	89.81	91.10	–
6	89.81	89.93	90.98	–
7	90.16	90.05	90.98	–
8	90.16	89.93	90.87	–
9	90.16	89.81	90.75	–

use publicly available code[1] that was developed for experiments in [5]. The code implements MV and EM algorithm [2] in conjunction with reputation Algorithm 2 (Hard penalty [5]). During each iteration, the reputation algorithm helps to exclude the volunteer with the highest penalty and recalculates the penalties for the remaining volunteers. We also benchmark IWMV [8]. Note that iterations in IWMV have different meanings. The accuracy of the compared algorithms on the test sample is presented in Table 2. Remarkably, IWMV converges right after 1 iteration with exactly the same predictions as MV. Though we observe a change of voters' weights (some are even flipped from 1 to −1), it does not influence the aggregated score of any image enough to alter the decision of MV. More surprisingly, all crowdsourcing algorithms perform on a par with MV. A possible explanation is the irregular task assignment leading, in particular, to a high percentage of images with only a few votes. To deal with this issue, we continue our analysis using image thresholding. Namely, we perform the same benchmarking for two subsets of the expert dataset. The subsets were obtained by filtering images with the number of votes less than the threshold; see Tables 3 and 4. Another possible explanation is that we mostly deal with reliable volunteers, and thus, crowdsourcing algorithms cannot profit from the detection of spammers or from flipping votes of malicious voters. To analyze the hypothesis, in the next subsection we classify volunteers using the annotator model proposed in [11].

---

[1]https://github.com/ashwin90/Penalty-based-clustering.

Table 3 Accuracy for 'crowdsourcing' algorithms with image thresholding

	MV	EM	Weighted MV	IWMV
Base	90.95	91.08	91.63	90.95
1	91.08	91.36	92.18	90.95
2	91.08	91.36	92.18	–
3	91.63	91.36	**92.32**	–
4	91.22	91.08	91.77	–
5	91.22	91.22	92.04	–
6	91.08	91.36	91.91	–
7	91.08	91.36	91.91	–
8	91.08	91.08	91.91	–
9	90.81	91.08	91.91	–

Only images with at least 4 votes are left in the expert dataset. In this case we have 729 images annotated by 1812 volunteers

Table 4 Accuracy for 'crowdsourcing' algorithms with image thresholding

	MV	EM	Weighted MV	IWMV
Base	94.55	94.55	95.05	94.55
1	94.55	94.55	95.05	94.55
2	94.55	94.55	95.05	–
3	94.55	94.55	95.05	–
4	94.55	94.55	95.05	–
5	94.55	94.55	95.05	–
6	94.55	94.80	95.30	–
7	94.55	94.80	95.30	–
8	94.55	94.80	95.30	–
9	94.80	94.80	**95.54**	–

Only images with at least 10 votes are left in the expert dataset. In this case we have 404 images annotated by 1777 volunteers

5.1 Annotator Models

As it was proposed in [11], a spammer is a person who labels randomly. A possible explanation is that a volunteer ignores images while labeling or does not comprehend the labeling criteria. '*More precisely an annotator is a spammer if the probability of an observed label is independent of the true label*' [11]. In what follows we define two important concepts, *the sensitivity* and *the specificity*. If the true label is 1, then the sensitivity α^j is defined as the probability that the volunteer j votes 1 (this probability corresponds to the true positive rate). If the true label is -1, then the specificity β^j is defined as the probability that the volunteer j votes -1. Note that volunteer j is a spammer if

$$\alpha^j + \beta^j = 1.$$

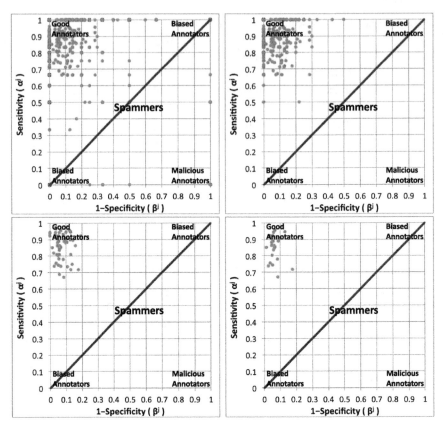

Fig. 4 We depict ROCs for volunteers having more votes than defined by a threshold. Threshold = 0 (the *upper left plot*), 12 (the *upper right plot*), 44 (the *lower left plot*), and 100. These thresholds leave 1813, 262, 52, and 24 volunteers, respectively. ROCs of spammers lie on the *red line*

This property suggests an easy way for detection of spammers. Namely, we simply depict the Receiver Operating Characteristic (ROC) plot containing details of individual performance; see Fig. 4. Since the task assignment was highly irregular, it is important to study how voting activity of volunteers influences the ROCs. Namely, Fig. 4 contains not one, but four ROCs, where each of them is obtained according to a different level of volunteer thresholding. This thresholding helps to remove volunteers that had a total number of votes less than that defined by the threshold. Figure 4 provides plausible observations for the dataset: there are no spammers among voters with more than 12 votes; good annotators prevail over all other types of annotators; we detect frequently voting volunteers (more than 100 votes) showing better accuracy than any examined algorithm, while there are no malicious voters. We conjecture that these are exactly the reasons why advanced algorithms (EM, IWMV) are on a par with MV.

6 Discussion

Comparing the results in Tables 1 and 2, it is remarkable that 'general purpose' learning algorithms slightly outperform 'special purpose' crowdsourcing algorithms. Moreover, the naïve heuristic (see Algorithm 1) based on reliability shows the best result. Also, numerical experiments show that MV performs on a par with all other algorithms (see Tables 1, 2, 3 and 4). The analysis of the ROCs of the volunteers suggests that surprisingly high accuracy of frequently voting volunteers coupled with the absence of spammers is a possible explanation for this result. The irregularity of volunteer-image assignment in the dataset with and a high percentage of images with a low number of votes may also contribute to this fact. Note that image thresholding by number of votes helps to improve the results of the 'crowdsourcing' algorithms (see Tables 2, 3 and 4) although the results are still on a par with MV. To summarize, good annotators and the irregularity eliminate any advantages of state of the art algorithms over MV. Thus, future research on the problem of aggregation of votes will benefit from a systems analysis approach capturing tradeoffs similar to the one shown in this paper.

Acknowledgments This research was supported by Russian Science Foundation, grant no. 14-11-00109, and the EU-FP7 funded ERC CrowdLand project, grant no. 617754.

References

1. Comber, A., Brunsdon, C., See, L., Fritz, S., McCallum, I.: Comparing expert and non-expert conceptualisations of the land: an analysis of crowdsourced land cover data. In: Spatial Information Theory, pp. 243–260. Springer (2013)
2. Dawid, A.P., Skene, A.M.: Maximum likelihood estimation of observer error-rates using the em algorithm. Appl. Stat. 20–28 (1979)
3. Dempster, A.P., et al.: Maximum likelihood from incomplete data via the EM algorithm. JRSS Ser. B 1–38 (1977)
4. Harvey, N.J., Ladner, R.E., Lovász, L., Tamir, T.: Semi-matchings for bipartite graphs and load balancing. In: Algorithms and Data Structures, pp. 294–306. Springer (2003)
5. Jagabathula, S., et al.: Reputation-based worker filtering in crowdsourcing. In: Advances in Neural Information Processing Systems. pp. 2492–2500 (2014)
6. Khattak, F.K., Salleb-Aouissi, A.: Improving crowd labeling through expert evaluation. In: 2012 AAAI Spring Symposium Series (2012)
7. Kim, H.C., Ghahramani, Z.: Bayesian classifier combination. In: International Conference on Artificial Intelligence and Statistics, pp. 619–627 (2012)
8. Li, H., Yu, B.: Error rate bounds and iterative weighted majority voting for crowdsourcing. arXiv:1411.4086 (2014)
9. Moreno, P.G., Teh, Y.W., Perez-Cruz, F., Artés-Rodríguez, A.: Bayesian nonparametric crowdsourcing. arXiv:1407.5017 (2014)
10. Pareek, H., Ravikumar, P.: Human boosting. In: Proceedings of the 30th International Conference on Machine Learning (ICML-13), pp. 338–346 (2013)
11. Raykar, V.C.: Eliminating spammers and ranking annotators for crowdsourced labeling tasks. JMLR **13**, 491–518 (2012)
12. Raykar, V.C., et al.: Learning from crowds. J. Mach. Learn. Res. **11**, 1297–1322 (2010)

13. Salk, C.F., Sturn, T., See, L., Fritz, S., Perger, C.: Assessing quality of volunteer crowdsourcing contributions: lessons from the cropland capture game. Int. J. Digital Earth 1–17 (2015)
14. See, L., et al.: Building a hybrid land cover map with crowdsourcing and geographically weighted regression. ISPRS J. Photogramm. Remote Sens. **103**, 48–56 (2015)
15. Simpson, E., et al.: Dynamic Bayesian combination of multiple imperfect classifiers. In: Decision Making and Imperfection, pp. 1–35. Springer (2013)
16. Tong, H., Li, M., Zhang, H., Zhang, C.: Blur detection for digital images using wavelet transform. In: 2004 IEEE International Conference on Multimedia and Expo, 2004. ICME'04, vol. 1, pp. 17–20. IEEE (2004)
17. Zauner, C.: Implementation and benchmarking of perceptual image hash functions. Ph.D. thesis (2010)
18. Zhu, X., et al.: Co-training as a human collaboration policy. In: AAAI (2011)

The Formal Models for the Socratic Method

Nico Huse and Nguyen-Thinh Le

Abstract In this paper, we present three formal models of three phases of the Socratic Method that is suggested by Nelson to be employed in teaching. The three phases are searching for examples, searching for attributes and generalizing the attributes. These formal models are intended to serve in a computerized learning environment where users can train with a chatbot to stimulate their critical thinking. This paper demonstrates the applicability and the usefulness of the formal models of the Socratic Method by showing an application that has been developed for group discussion where the chatbot acts as a discussion leader who applies the Socratic Method. The contribution of this paper is twofold. First, in the dialogue models, we integrated critical questions using the question taxonomy of Paul and Elder in the three phases of the Socratic Method. Second, the formalization of the three phases of the Socratic Method using state diagrams is a new innovation.

Keywords Socratic questioning · Socratic discussion · Critical thinking

1 Introduction

Asking questions is an important skill that is required in many institutional settings, e.g., interviews conducted by journalists [6], medical settings [7], courtrooms [1]. For teachers, asking questions is almost an indispensable teaching technique. Dillon [8] investigated questions generated by teachers in 27 upper classrooms in six secondary schools in the USA and reported that questions accounted for over 60 % of the teachers' talk. The benefits of using questions in instruction are multi-faceted and have been reported in many research studies [13, 14, 22]. Not only teachers'

N. Huse · N.-T. Le (✉)
Department of Informatics, Humboldt-Universität zu Berlin, Berlin, Germany
e-mail: nguyen-thinh.le@hu-berlin.de

N. Huse
e-mail: nico.huse@googlemail.com

© Springer International Publishing Switzerland 2016
T.B. Nguyen et al. (eds.), *Advanced Computational Methods for Knowledge Engineering*, Advances in Intelligent Systems and Computing 453, DOI 10.1007/978-3-319-38884-7_14

questions can enhance learning, but also students' question asking can benefit learning. The evidence from research studies provides a solid empirical basis to support the inclusion of students' question asking in teaching in order to enhance comprehension [21], cognitive and metacognitive strategies use [23], and problem-solving abilities [3] of students. Researchers suggested that teachers should pose questions that encourage higher-level thinking of students because they need to be familiarized with different levels of thinking and to use knowledge of the lower-level productively [5]. In addition, Morgan and Saxton [14] demonstrated that well-chosen higher-order questions can not only be used to assess student's knowledge but also to extend his/her knowledge, to improve his/her skills of comprehension and application of facts and also to develop his/her higher-order thinking skills. Yet the evidence is that the majority of questions teachers use in their classrooms in order to check knowledge and understanding, to recall of facts or to diagnose student's difficulties [5], and only about 10 % of questions are used to encourage students to think [4].

Socratic dialogue is considered an effective teaching approach that is used to stimulate critical thinking [18]. Paul defined "critical thinking" as: "Thinking explicitly aimed at well-founded judgement, utilizing appropriate evaluative standards in an attempt to determine the true worth, merit, or value of something." Based on the definition of "critical thinking" Paul argued that "critical thinking provides us with definitive and specific tools". He linked the relationship between critical thinking and the Socratic Method as follows: "Critical thinking [...] is the key to Socratic questioning because it makes the intellectual moves used in Socratic dialogue explicit and accessible to anyone interested in learning it, and willing to practice it." There exists a huge body of literature about Socratic dialogues and the Socratic teaching approach. However, the research gap is how to formalize the process of Socratic teaching. In this paper, we present our attempt to formalize the three steps of the Socratic Method: searching for examples, searching for attributes, and summarization.

2 State of the Art of Computer Modeling of Critical Thinking

2.1 The Socratic Method

The Socratic Method is rooted back to the philosopher Socrates about whom we know from the books of Platon, one of Socrates' disciplines. "He [Socrates] is best known for his association with the Socratic method of question and answer, his claim that he was ignorant (or aware of his own absence of knowledge)".[1] It may be true that Socrates was highly skilled at questioning. However, it is not easy to

---

[1]http://www.iep.utm.edu/socrates/.

emulate the types of questions he asked at any given point in a discussion. By studying dialogues of Socrates, Paul and Elder [18] explicated the components and processes underlying Socrates' dialogues. They identified a classification of six classes of Socratic questions: (1) questions that require clarification, (2) questions probing assumptions, (3) questions probing reasoning and evidence, (4) questions probing perspective, (5) questions probing implications, (6) questions about the question. This question taxonomy provides us with specific tools for critical questioning.

Nelson [15] analyzed the questioning technique of Socrates and developed the so-called Socratic Method. A Socratic dialogue, according to Nelson, starts with a self-experienced example. Nelson identified in the habit of Socrates' dialogues that Socrates used observations of daily life as examples and pre-conditions of certain judgment to lead the dialogue partner to the less certain judgment. This is referred to as the abstraction process. Horster [11] investigated the theoretical assumptions of the Socratic Method, modified the abstraction process proposed by Nelson and extracted the Socratic dialogue in the following steps.

Given a discussion topic (e.g., freedom, happiness, sense of life, etc.), the first step is to ask the dialogue partner to give self-experienced examples for the specified topic. The attributes and features of the topic should be contained in each example (i.e., to be sure that the example is relevant to the topic). The second step is to collect those attributes and features. As the third step, the collected attributes and features will be summarized. These three steps of the Socratic Method have been being applied widely not only in dialogues, but also in group discussion in the sense of Socratic teaching.

2.2 Applications that Support the Socratic Method

Several educational applications support tutorial dialogues. Olney and colleagues [16] presented a method for generating questions for tutorial dialogue. This involves automatically extracting concept maps from textbooks in the domain of Biology. Five question categories were deployed: hint, prompt, forced choice question, contextual verification question, and causal chain questions. Also with the intention of supporting students using conversational dialogues, Person and Graesser [19] developed an intelligent tutoring system that improves students' knowledge in the areas of computer literacy and Newtonian physics using an animated agent that is able to ask a series of deep reasoning questions according to the question taxonomy proposed by Graesser and Person [9]. Lane and VanLehn [12] developed PROPL, a tutor, which helps students build a natural-language style pseudo-code solution to a given problem. All these educational applications deployed some kinds of dialogues, however, they did not apply the Socratic Method.

Hoeksema [10] developed a group discussion environment that was intended to serve virtual Socratic dialogues. The author used the collaborative learning environment Cool Modes ("Collaborative Open Learning Modelling and Designing System", Pinkwart et al. [20]). The Socratic dialogues using this discussion environment were intended to be held similarly in a usual face-to-face environment. Whereas this work focused on developing an environment for Socratic group discussions, our goal is to formalize the Socratic Method in order to help students develop critical thinking.

Otero and Graesser [17] developed a computational mechanism for triggering questions. Based on the computational model of a given topic text, the computer should be able to decide for which situation which question can be posed to the student. For this purpose, a formalism for describing conditions of posing questions is required. Otero and Graesser [17] proposed to use production rules. These rules specify which particular questions should be generated when particular elements or configurations of information in the text occur. Each production rule is an "if state, then action" expression. If a particular state (or Boolean configuration of states) exists, then an action or action sequence is performed [2].

The goal of our work is to formalize the dialogue steps of the Socratic Method. Whereas Otero and Graesser used production rules as a formalism, we propose to apply state diagrams, because the dialogue steps could be mapped to the dialogue states intuitively. To our best knowledge, no formalization for the Socratic Method has been developed yet. The formalization of the Socratic Method for dialogues described in this paper contributes both to the Socratic teaching community (through the integration of critical questions of Paul and Elder in the Socratic Method) and the community of educational technologies.

3 The Formal Models of the Socratic Method

In this section, we present the formal models for the Socratic Method that was suggested by Nelson [15] and extended by Horster [11] and is applied in dialogues or in group discussions. In the developed models, a computer agent is required in order to start conversation by imitating the role of the discussion leader. The computer agent is called "chatbot" in the developed models who attempts to apply the Socratic Method in order to stimulate the discussion participants to critically think about the discussion topic by searching for examples, searching for attributes, and searching for generalized attributes. The dialogue starts when the chatbot explains the phases of the discussion (search for examples, search for attributes, and summarizing the attributes) and gives a discussion topic.

The first phase starts when the chatbot asks all participants to give examples (cf. Fig. 1). After a participant Tx gives an example, the chatbot asks Tx to explain more about the example by applying the question class "clarification". At this time, Tx thinks about the given example again and may dismiss it or elaborate on it. As the next step, the chatbot asks other participants if they have any concern about the example given by Tx. If any questions arise, Tx needs to explain his/her example again. Otherwise, the given example can be added to the collection of examples for

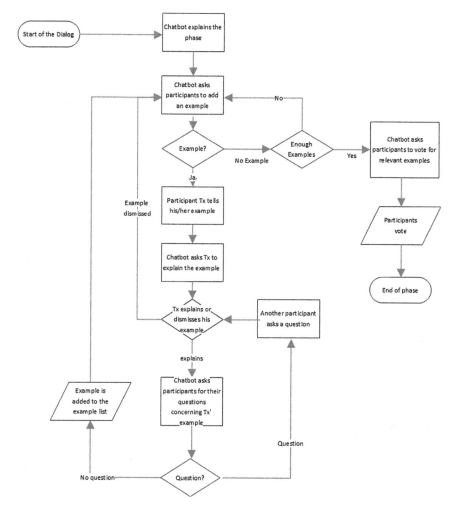

Fig. 1 Search for an example

the discussion topic and this collection will be shown to other participants. If the example collection is enough (the chatbot may determine the threshold for "enough", for instance, the threshold is equal the number of participants minus 1), the chatbot asks the participants to vote the best example that will be worked out in the next phase (i.e., searching for attributes). After this vote, the first dialogue phase is finished.

The best voted example is used in the second phase to elicit relevant attributes. First, the chatbot repeats all collected examples (cf. Fig. 2). The chatbot indicates the list of elicited attributes as if the discussion leader uses a whiteboard or flipchart to collect attributes given by the discussion participants (assuming that at the beginning, the flipchart is empty). Then, the chatbot asks the participants to elicit attributes

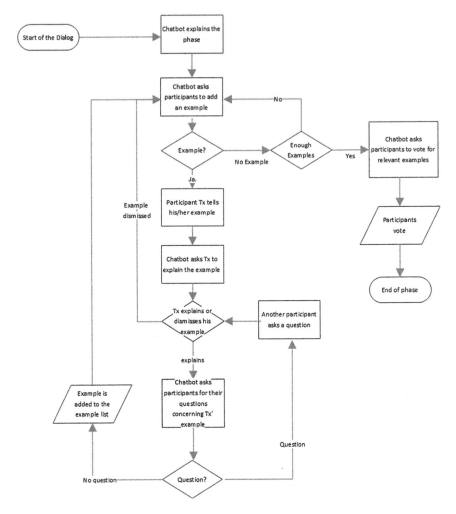

Fig. 2 Search for attributes

from the examples. If a participant Tx gives feedback (i.e., he/she wants to name an attribute), the chatbot asks Tx to explain the attribute using the question class "clarification" (cf. Paul and Elder [18]). After Tx has elaborated on the named attribute, the chatbot asks other participants about their opinion or whether they have any question concerning the named attribute. If any participant Ty has a question, the chatbot asks him/her about his/her question using the class "questions about the question". If any participant Ty has an opinion, the chatbot asks about his/her opinion using the class "questions about viewpoints and perspectives". The question (of the class "questions about the question" or of the class "questions about viewpoints and perspectives") scrutinize the question/opinion of Ty and help him/her to think about his/her question/opinion again. In case, the participant Ty has a question, the chatbot forwards this question to the participant Tx and requests him/her to answer the Ty's question. This cycle is repeated until no questions from other participants arise. In case, the participant Ty has an opinion, the chatbot forward this opinion to Tx. Tx may dismiss his/her named attribute, change it or remain with that attribute. If neither questions nor opinions arise, the named attributed of the participant Tx will be added to the list of collected attributes. The chatbot decides to finish this discussion phase if the list of collected attributes is "enough" (e.g., the threshold can be determined by the number of participants minus 1). Similar to the threshold for searching examples, here, the chatbot may determine that the threshold for attributes is equal the number of participants minus 1.

Similar to the previous two phases, at the beginning of the third phase (cf. Fig. 3), the chatbot takes the role of the discussion leader and explains the rules of this phase. Then, the chatbot presents the list of collected attributes for the discussion topic and asks the participants to generalize two or more attributes. If a participant Tx notifies that he/she wants to generalize the attributes, the chatbot asks

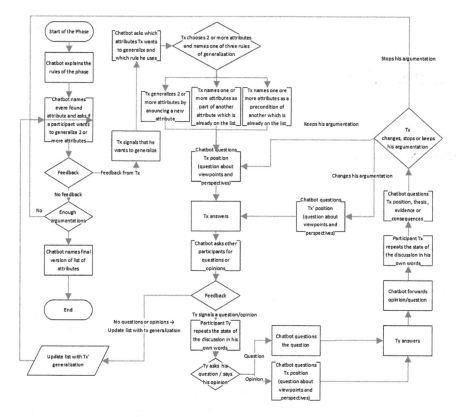

Fig. 3 Generalizing the attributes

him/her which attributes on the list he/she wants to generalize and which rule he/she wants to apply. Horster [11] suggested three ways of summarizing the attributes: (1) an attribute is more general than another, (2) a hypernym is used for two or several attributes, and (3) an attribute is a pre-condition for another attribute. The participant Tx will have the option to select two or more attributes from the attribute list and select one of the three generalization rules. After Tx has selected a generalization rule, the chatbot will apply the class of "questions that probe reason and evidence" to ask the participant Tx to explain his/her decision.

Then, the chatbot asks other participants about their opinions or whether they have any questions. If a participant Ty notifies that he has a question, first he needs to summarize the status of the discussion in his/her own words, and then asks a question or gives his/her opinion. If the participant Ty asks a question, the chatbot asks about his/her question using the class "questions about the question". If the participant gives an opinion, the chatbot asks a question about this opinion using the question classes "questions that probe reason and evidence". After the participant Ty answers the question of chatbot, his/her answer is forwarded to the participant Tx. Then, the chatbot asks the participant Tx to summarize the state of the discussion in his/her own words. Then, the chatbot asks a question about his/her position. Upon on this question, the participant Tx may withdraw his/her argument for generalization of attributes, change or remain the argument (for generalizing the attributes). In case, all other participants agree with the argument for the generalization of attributes (i.e., no question, no opinion), the chatbot updates the list of generalized attributes. If the list of generalized attributes is "enough", the chatbot can stop the discussion. Horster [11] proposed an extension of the Socratic Method that the discussion leader goes back to the phase 2 (searching for attributes) and take the next best example to be discussed in the phases 2 and 3.

4 Implementation

In order to verify the applicability of the formal models for the three phases of the Socratic Method, we have decided for an application that is designed to support group discussion. The application provides a chatbot who acts as a discussion leader applying the Socratic Method to stimulate the discussion participants thinking about a discussion topic. In order to show the scalability of the three models of the Socratic Method, in this group discussion scenario, the number of participants will be more than two. The architecture of the application consists of three modules: the user interface, the message handling module, and the chatbot module.

Since the intended application should support group discussion, we decided to use WhatsApp Messenger, a social media frontend that is being used widely by more than one billion users.[2] Many people are familiar with the user interface of WhatsApp. Thus, users can focus on the discussion. Most users have their own Smart phones or tablets and thus, can join the group discussion from everywhere.

In order to communicate with a group of users, we need an interface to the server of WhatsApp. The message handling module is implemented using Yowsup[3] (a Python program), which is responsible for sending and receiving messages. Yowsup receives messages from each user and forward them to all discussion participants. Yowsup imitates to WhatsApp as a normal user that requires a user ID and a password. Each WhatsApp account is associated with a telephone number. Using this number, the user ID and a password will be generated. In order to get a WhatsApp account for the chatbot, we use the following Yowsup commado "yowsup-cli" that sends an account request for the telephone number 4930XXXXXXX: `yowsup-cli registration -requestcode voice -phone 4930XXXXXXX -cc 49 -mcc 123 -mnc 456`.

In order to send messages using Yowsup, we use the commando tool: `yowsup-cli demos -l 4930XXXXXXX:XXXPASSWORDXXX -s Receiver TEXT`. As receiver of a message, we need to fill the placeholder "Receiver" with a complete telephone number with country code. For a group of receivers, instead using the telephone number of each receiver, we need a group ID that is created by WhatsApp.

In order to receive messages, we modified the echo-client that is provided by Yowsup so that WhatsApp only gets the latest available message. The original version of the echo-client runs permanently after each message sending commando has been initiated.

The chatbot module, which is the main module of the application, implements the developed formal models of the Socratic Method (cf. Sect. 3). This module controls the discussion phases. The tasks of the chatbot include explaining and observing the phases of the dialog and its rules as a discussion leader for a Socratic group discussion would do. The chatbot controls and drives the conversation. It receives messages from Yowsup and makes sure that only active participants are allowed to participate in the discussion. It manages the list of examples in the phase of searching examples, the list of attributes in the phase of searching attributes and the list of generalized attributes in the generalization phase. The chatbot chooses questions based on the state of the dialogue model. For each state in the diagrams of the dialogue models, the chatbot has a set of questions it can ask. These questions are mostly chosen by random choice. The chatbot mostly ignores to understand the context of user's input. Only in some situations, it tries to identify specific word or even groups of words by checking the input against regular expressions. These are situations where the Chatbots needs a "yes" or "no" answer from a participant. For example, the adjectives "free", "relaxed", "calm", "happy" are specific for the topic

---

[2]http://blog.whatsapp.com/615/WhatsApp-kostenlos-und-n%C3%BCtzlicher-machen   (Access: 07/03/2016).

[3]https://github.com/tgalal/yowsup/wiki.

"happiness". In case, the users use one of these adjectives in their answers, the chatbot will pose a corresponding question: "What do you mean with free?", "Why were you so relaxed?", "Why were you so calm?", "Why did you fell happy?" The authors tried to use this technique in the first phase of the dialogue (the phase of collecting examples) to catch specific adjectives which we thought are specific for the topic of the dialogue. If the chatbot detects one of the topic-specific adjectives, there would be a chance that the chatbot asks a pre-specified question. In addition to the task of controlling the conversation, another task of the chatbot is archiving the history of discussion.

5   A Case Study: A Socratic Group Discussion Application

The app has been used by six users. The chatbot initiates the discussion by giving the topic "What is happiness?" In the first phase (searching for examples) (cf. Fig. 4), the chatbot collects examples for the topic "happiness" and update the list of examples (the list represents a whiteboard or a clip chart for a discussion leader to take notes). The colored marks were intentionally used to anonymize the name of the users. Figure 4 demonstrates how the chatbot acts as a discussion leader applying the Socratic Method to help users think about examples for the given discussion topic.

After finishing the first phase (i.e., the chatbot has collected five examples, because in this scenario we have six users), the chatbot initiates the phase of searching attributes (cf. Fig. 5).

Fig. 4 A demonstration of phase "searching for examples" of the Socratic group discussion

After collecting five attributes for the topic "happiness", the chatbot asked the users to generalize the attributes (cf. Fig. 6). On the left side of this figure, a participant generalized the attributes "love" and "satisfaction".

These three figures (Figs. 4, 5 and 6) show only six screenshots of the discussion of six participants that participate in the evaluation study within about 2 h. After the discussion session, we asked the users to answer a questionnaire in order to evaluate the usefulness of the models for Socratic Method in general and to evaluate the specific questions/dialogue moves initiated by the Socratic chatbot. After this evaluation, we can learn that the questions asked by the Chatbot sometimes were not in the context of the discussion. During the course of the dialogues, some

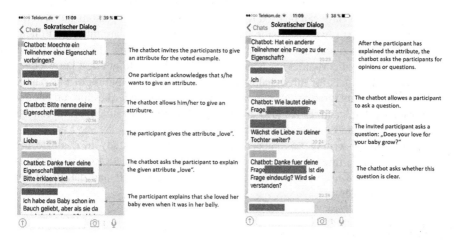

Fig. 5 A demonstration of phase "searching for attributes" of the Socratic group discussion

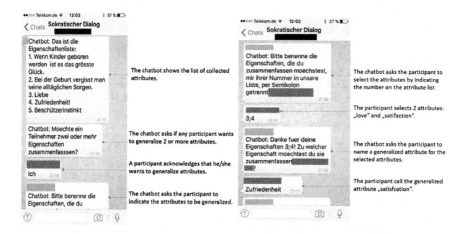

Fig. 6 A demonstration of phase "generalizing the attributes" of the Socratic group discussion

participants started to take the Chatbot less serious. Another fact we have learned was that the Chatbot was very rigid in controlling the dialogues, which led to fact that there was no real discussion between the participants. Due to space limit, the detailed results of the evaluation will be published in another article.

6 Conclusions and Future Work

In this paper, we have presented three models for three dialogue phases of the Socratic Method by applying state diagrams. These models have been validated in a group discussion scenario using the WhatsApp frontend. This validation showed that the models not only can be used for dialogues between two persons but also for a group of more than two discussion participants. The models are intended to help students develop critical thinking for a given discussion topic. Since no formal model for the Socratic Method has been proposed until now, to our best knowledge, the introduced models for three dialogue phases are the contribution of this paper. In addition, the integration of critical questions using the question taxonomy developed by Paul and Elder [18] in the steps of the Socratic Method can be considered a second contribution in the paper.

At the time of writing this paper, we are conducting a study to evaluate the usefulness of these three dialogue models. We will analyze data from a questionnaire regarding the usefulness of the Socratic dialogue models and data from the discussion protocols.

References

1. Atkinson, J.M., Drew, P.: Order in Court. Macmillan, London (1979)
2. Anderson, J.R., Corbett, A.T., Koedinger, K.R., Pelletier, R.: Cognitive tutors: lessons learned. J. Learn. Sci. **4**, 167–207 (1995)
3. Barlow, A., Cates, J.M.: The impact of problem posing on elementary teachers' beliefs about mathematics and mathematics teaching. Sch. Sci. Math. **106**(2), 64–73 (2006)
4. Brown, G., Wragg, E.C.: Questioning. Routledge (1993)
5. Chafi, M.E., Elkhouzai, E.: Classroom interaction: investigating the forms and functions of teacher questions in Moroccan Primary School. J. Innov. Appl. Stud. **6**(3), 352–361 (2014)
6. Clayman, S., Heritage, J.: The News Interview: Journalists and Public Figures on the Air. Cambridge University Press, New York (2002)
7. Drew, P., Heritage, J.: Analyzing talk at work: an Introduction. In: Drew, P., Heritage, J. (eds.) Talk at Work: Interaction in Institutional Settings, pp. 3–65. Cambridge University Press, New York (1992)
8. Dillon, J.T.: Questioning and Teaching. A Manual of Practice. Croom Helm (1988)
9. Graesser, A.C., Person, N.K.: Question asking during tutoring. Am. Educ. Res. J. **31**(1), 104–137 (1994)
10. Hoeksema, K.: Virtuelle Sokratische Gespräche - Umsetzung einer Idee aus dem Philosophieunterricht. In: Proceedings on "Modellierung als Schlüsselkonzept in intelligenten Lehr-/Lernsystemen" (2004)

11. Horster, D.: Das Sokratische Gespräch in Theorie und Praxis. Opladen (1994)
12. Lane, H.C., Vanlehn, K.: Teaching the tacit knowledge of programming to novices with natural language tutoring. J. Comput. Sci. Edu. **15**, 183–201 (2005)
13. Lin, L., Atkinson, R.K., Savenye, W.C., Nelson, B.C.: Effects of visual cues and self-explanation prompts: empirical evidence in a multimedia environment. Interac. Learn. Environ. J. (2014)
14. Morgan, N., Saxton, J.: Asking Better Questions. Pembroke Publishers, Makhma, ON (2006)
15. Nelson, L.: Die sokratische Methode. Gesammelte Schriften in neun Bänden, Band 1, Hamburg. Publisher Paul Bernays u. a. Meiner (1970)
16. Olney, A.M., Graesser, A., Person, N.K.: Question generation from concept maps. Dialogue Discourse **3**(2), 75–99 (2012)
17. Otero, J., Graesser, A.C.: PREG: elements of a model of question asking. Cogn. Instr. **19**(2), 143–175. Lawrence Erlbaum Associates (2001)
18. Paul, R., Elder, L.: Critical thinking: the art of socratic questioning, Part I. J. Dev. Educ. **31**(1), 36–37 (2007)
19. Person, N.K., Graesser, A.C.: Human or Computer? AutoTutor in a bystander turing test. In: Cerri, S.A., Gouardères, G., Paraguaçu, F. (eds.) Proceedings of the 6th International Conference on Intelligent Tutoring Systems, pp. 821–830. Springer (2002)
20. Pinkwart, N., Hoppe, U., Gaßner, K.: Integration of domain-specific elements into visual language based collaborative environments. In: Proceedings of 7th International Workshop on Groupware. IEEE Computer Society, Los Alamitos, California, pp. 142–147 (2001)
21. Rothstein, D., Santana, L.:. Teaching students to ask their own questions. Havard Educ. Lett. **27**(5) (2014)
22. Tenenberg, J., Murphy, L.: Knowing what I know: an investigation of undergraduate knowledge and self-knowledge of data structures. Comput. Sci. Educ. **15**(4), 297–315 (2005)
23. Yu, F.Y., Pan, K.J.: The effects of student question-generation with online prompts on learning. Educ. Technol. Soc. **17**(3) (2014)

Using Local Weather and Geographical Information to Predict Cholera Outbreaks in Hanoi, Vietnam

Nguyen Hai Chau and Le Thi Ngoc Anh

Abstract In 2007, repeated outbreaks of cholera in Hanoi have raised the need to have up-to-date evidence on the impact of factors on cholera epidemic, which is essential for developing an early warning system. We have successfully built models to predict cholera outbreaks in Hanoi from 2001 to 2012 using Random Forests method. We found that geographical factors—the number of cholera cases of a district of interest and its neighbours—are very important to predict accurately cholera cases besides the weather factors. Among weather factors, temperature and relative humidity are the most important. We also found that prediction accuracy of our models, measured in adjusted coefficient of determination, will decrease by 0.0076 if prediction length increases by one day.

Keywords Cholera outbreaks prediction · Random forests · Geographical information · Time series

1 Introduction

Cholera is a global public health issue despite the decrease of morbidity and mortality in recent years [1, 2]. Cholera is an acute watery diarrhea caused by a multiplication of gram-negative toxigenic bacterium namely Vibrio Cholera (V. Cholera) in human intestine [3, 4]. It is estimated that approximately 200 serotypes of V. Cholera are available, of which O1 and O139 are the primary cause of cholera epidemics and endemics worldwide [2, 4]. Cholera is regularly considered in the relations with unclean water and poor sanitation infrastructures, especially in low and

N.H. Chau (✉)
Faculty of Information Technology,
VNUH University of Engineering and Technology, Hanoi, Vietnam
e-mail: chaunh@vnu.edu.vn

L.T. Ngoc Anh
Information Technology Department, Hanoi Medical University, Hanoi, Vietnam
e-mail: lengocanh@hmu.edu.vn

© Springer International Publishing Switzerland 2016
T.B. Nguyen et al. (eds.), *Advanced Computational Methods
for Knowledge Engineering*, Advances in Intelligent Systems
and Computing 453, DOI 10.1007/978-3-319-38884-7_15

195

middle-income countries [4, 5]. Annually, about 2.9 million cases and 91,000 deaths occur as a consequence of cholera infection.

Along with water and hygiene status, previous studies demonstrated that climate variability partly contributes to the widespread of V. Cholera [6]. For example, studies in Africa indicated that the increase of temperature and rainfall results in the rise of cholera cases [7, 8]. Furthermore, studies in Bangladesh showed that temperature and sunshine hours might relate with cholera occurrence [9]. In a recent report, the World Health Organization has underlined that climate variables have the central role on the temporal and spatial distribution of infectious diseases, raising the need to develop an early warning system based on meteorological factors [10, 11]. Therefore, establishing climate-based predictive models for cholera epidemic are necessary for prompt prevention and intervention in the longer run.

Vietnam experienced cholera epidemics in the twentieth century, especially in 1960s and 1990s, and most of cases were reported in Southern regions [2]. However, in 2007 and 2008, the cholera outbreaks occurred with the majority of affected provinces in Northern regions, including Hanoi [3, 12, 13]. Until April 2008, there were 3,271 cases reported from 18 provinces [3]. The reasons for this epidemic were argued not to rest only with water or food contamination [12]. Therefore, understanding the association between cholera cases and other factors as climate variability is necessary to develop the strategies to control, monitor and prevent the cholera outbreaks.

2   Related Works

Ali et al. [14] studied cholera data of Matlab, Bangladesh from 1988 to 2001 and concluded that the number of cholera cases in the study area is strongly associated with local temperature and sea surface temperature (SST). Time series analysis is the method used in this research.

Reiner et al. [15] successfully build a model that is able to predict the number of cholera cases in Matlab, Bangladesh 11 months in advance. Data sets used in this research are local weather, southern oscillation index (SOI) and flooding condition from 1995 to 2008. As a result from the research, SOI and flooding condition are main positive factors to the number of cholera cases in Matlab. Prediction method used in this research is simulation using multidimensional inhomogeneous Markov chain (MDIMC).

Min et al. [16] used MaxEnt model, a maximum expectation like model, to analyze China's cholera outbreaks from 2001 to 2008. As their results, precipitation, temperature and the location's altitude are strongly linked to the number of cholera cases. Distance of the location to the sea coast, relative humidity and atmospheric pressure are also linked to the number of cholera cases. The sun hours and river height discharge are independent to the number of cholera cases.

In another research, Min et al. [17] used satellite and geographical data to find influences of SST, sea surface height (SSH) and ocean chlorophyll concentration (OCC) to the number of cholera cases in China from 1999 to 2008. Results show that changes in SST and SSH are associated immediately to the number of cholera cases while OCC has 1 month lag effect.

Emch et al. [18] studied effects of local weather parameters to the number of cholera cases in Matlab, Bangladesh from 1983 to 2003 and in Nha Trang, Hue (Vietnam) from 1985 to 2003. Results show that high OCC have positive association to the number of cholera cases in Matlab, increasing SST correlates with the number of cholera cases in Hue and increasing of river height correlates with the number of cholera cases in Nha Trang. In the research, the authors show that local weather has 2 months lag effect. They use univariate and multivariate statistical analysis methods.

The above works show that local weather parameters such as temperature, relative humidity, SOI, SST, SSH have different association to number of cholera cases in different areas.

In Vietnam, several previous studies mentioned the association of local environment with occurrence of cholera cases. A study of Kelly-hope et al. in Vietnam suggested the significant link of precipitation and cholera outbreaks at 0-month lag during 1991–2001 [5]. Emch et al. found that the significant predictors of cholera infected included increasing SST and river height [6]. However, the outbreak of cholera from 2007 to 2009 in Hanoi has raised the need to have more reliable evidence about the impact of climate factors along with traditional environment indicators. In order to provide more comprehensive and up-to-date evidence, this paper aimed to investigate the relationships of cholera incidence with weather, geographical factors and the SOI climate change indicator in Hanoi during 2001–2012.

The rest of this paper is organized as follows. In Sect. 3, we describe Hanoi local information and data sets used for research. Prediction models building is presented in Sect. 4. We analyze prediction results in Sect. 5 and concludes the paper in Sect. 6.

3 Description of the Study Area and Data Sets

To build prediction models of cholera prediction in Hanoi, Vietnam, we use the following data sets: Cholera cases, local weather, geographical data of Hanoi and SOI data set. In the following we describe Hanoi local information and the data sets.

3.1 Study Area

Hanoi, located at 21°01′42.5″N, 105°51′15.0″E, is the capital and the second largest city of Vietnam. Its population in 2009 is approximately 2.6 millions in urban districts and 6.5 millions in metropolitan areas. From 2001 to 2012, Hanoi is divided into 11 urban districts, a district-level town and 17 metropolitan districts. The

Fig. 1 Hanoi administrative map. The *black circle* indicates location of Lang weather station

districts and the town are further subdivided into more than five hundreds communes, wards and commune-level town. For simplicity, in this research we refer administrative level 2 as "district", level 3 as "commune" assuming that Hanoi's administrative level is 1.

Hanoi is located in the northern region of Vietnam. It is embraced by the Red River and roughly 100 km far from coastal area. Hanoi's climate is warm humid subtropical, identified as *Cwa* in Köppen climate classification system. It has four distinct seasons: spring (February–April), summer (May–August), fall (September–October) and winter (November–January).

3.2   Geographical Data Sets

The data set contains administrative boundaries of districts and communes, roads, rivers and water areas at 1:50,000 scale. Map of Hanoi and its 29 districts is in Fig. 1.

3.3   Cholera Data Set

The data set contains all observed cholera cases of Hanoi from Jan 01, 2001 to Dec 31, 2012. Each record of the data set contains patient's name, age, sex, date of infection and his/her home address at least to commune administrative level. We aggregate

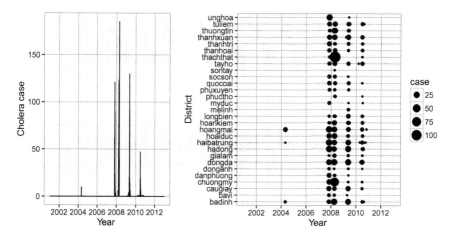

Fig. 2 Cholera cases of Hanoi from 2001 to 2012. *Left* Daily cholera cases of Hanoi. *Right* daily cholera cases by district; size of *black circles* is proportional to the number of cholera cases

the data set to calculate the number of daily cholera case per district. Daily aggregated cholera cases of Hanoi and daily aggregated cholera cases per district are in Fig. 2. The data set shows that there were five cholera outbreaks in Hanoi in 2004, 2007, 2008, 2009 and 2010. The other years are cholera-free.

3.4   Local Weather Data Set

This data set contains daily relative humidity (min, max and mean), daily temperature (min, max and mean), daily sun hours, daily wind speed and daily precipitation measured by Lang weather station in Hanoi. The Lang weather station locates in Cau Giay district of Hanoi as shown in Fig. 1 and its measurements are representative for Hanoi weather. Figure 3. illustrates the local weather data set.

3.5   SOI Data Set

We use SOI data collected from a website of Queensland government, Australia [19]. The data set contains daily SOI measurement from 1991 to the day of getting data. Daily SOI data from 2001 to 2012 is plotted in Fig. 4.

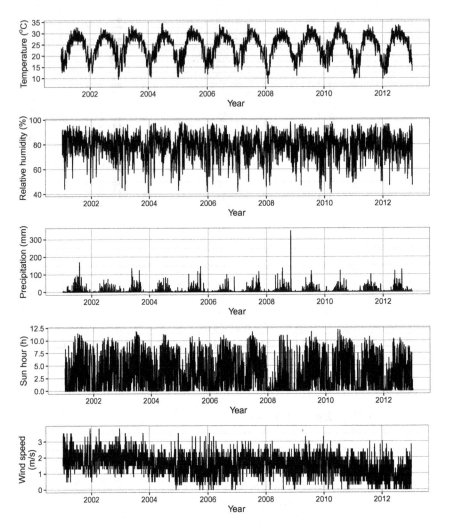

Fig. 3 Daily local weather data of Hanoi from 2001 to 2012 as measured by Lang weather station. From *top* to *bottom*: Mean temperature, mean relative humidity, precipitation, sun hour and wind speed

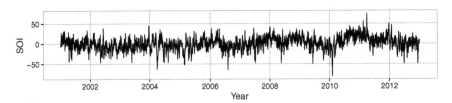

Fig. 4 Daily SOI from 2001 to 2012

4   Features Selection and Models Building

4.1   Features Selection

In previous works of cholera prediction, researchers often use monthly aggregated data as their main data sets. We use daily aggregated data. The reasons include sparsity and the number of data points in our monthly aggregated data set. If we aggregate the cholera and weather data sets by month, we have 144 observations among which 19 are with cholera. In addition, the maximum length of each cholera outbreak of Hanoi in month is only three, making "positive" data series very short and difficult to predict.

We aggregate data sets described in Sect. 3, except the geographical data set, by day and mix them into one. We consequently have a final data set, denoted as FS, with 35 variables and 4383 observations. Six of the variables are weather ones including temperature, relative humidity, precipitation, sun hour, wind speed and SOI. The others are daily cholera cases of 29 districts. Figure 5 shows a correlogram of the FS data set. It is obvious that the weather variables have very close to zero correlation

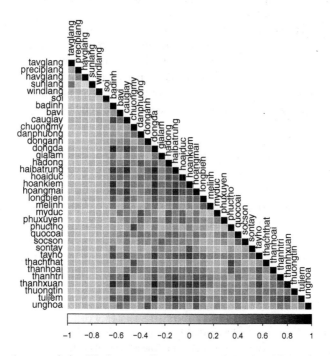

Fig. 5 Correlogram of the FS data set. Temperature, relative humidity, precipitation, sun hour, wind speed and SOI variables are denoted as `tavglang`, `havglang`, `preciplang`, `sunlang`, `windlang` and `soi`, respectively. Other variables, named after districts names, are daily cholera cases corresponding to districts

with the cholera case variables. However the cholera case variables are correlated. Cholera case variables of geographically closed districts are generally more correlated. This fact suggests us to use subsets of cholera case variables as additional predictors in combination with weather variables for building prediction models. However, the cholera case variables of districts are also outcomes. Therefore, we can only use the past values of cholera variables for prediction. In the following, we describe the building of our prediction models.

4.2 Models Building

While other studies on cholera outbreak prediction mainly consider study areas as atomic, we use districts of Hanoi as geographical units. For each district of Hanoi, we build three predictive models namely complete, weather-independent and geographical-independent, abbreviated as CP, WI and GI, respectively. Our purpose is twofold. Firstly, we want to choose the best model for all districts. Secondly, we want to evaluate effects of geographical and weather variables to the accuracy of prediction. Table 1 explains the predictor groups of each model. All the models has the number of cholera case as an outcome.

Each model has a lag parameter l measured in day. The parameter means that we use the number of cholera cases of the current and previous $l - 1$ days in a district of interest as a predictor for the model. It also means that we predict the number of cholera case of the district in next l days. For each model, we also use the past number of cholera cases of all neighbours of the district of interest and past weather information as additional predictors. Two districts are neighbours if they share parts

Table 1 Description of predictors for CL, WI and GI models

Predictor group	Model		
	CP	WI	GI
Weather information	Mean temperature	*Not used*	Mean temperature
	Mean relative humidity		Mean relative humidity
	Precipitation		Precipitation
	Sun hours		Sun hours
	Wind speed		Wind speed
	SOI		SOI
Geographical information	The number of cholera cases of a district D	The number of cholera cases of a district D	*Not used*
	The number of cholera cases of D's neighbour districts	The number of cholera cases of D's neighbour districts	

Table 2 An example of sliding window training and testing for the CP model

w_1	w_2	w_3	w_4	w_5	w_6	w_7	w_8
d_4	d_5	d_6	d_7	d_8	d_9	d_{10}	d_{11}
n_1	n_2	n_3	n_4	n_5	n_6	n_7	n_8
Training data set 1			Testing data set 1				
	Training data set 2			Testing data set 2			
		Training data set 3			Testing data set 3		

In this example, lag parameter l is 3

of their administrative borders. Using SQL spatial queries in PostgreSQL/PostGIS we easily define neighbours of each district in Hanoi.

Instead of using statistical techniques for model building in many other studies, we adopt a machine learning approach. After a process of try and error, we found that the Random Forests (RF) regression method is the most suitable for our prediction models. RF regression is a supervised learning method. It learns on training data sets and predict on testing ones. Since all variables of the FS data set is time series, we apply the rolling forecasting origin techniques by Hyndman and Athanasopoulos [20]. Using this technique, we first create an initial window that has s_1 data points as the first training data set. The testing data set is next s_2 data point. Note that each data point in the training data set contains all predictors and the outcome, and each data point in the testing data set contains only predictors. The window is then shifting along the time axis until no data point left. The supervised model is built during the shifting and improved along the time axis. We set $s_1 = s_2 = l$ in all models.

Table 2 illustrates training and testing data sets of a CP model with lag parameter $l = 3$ of a district D. In the table, $w_1, w_2, ..., w_8$ are values of weather variables, $n_1, n_2, ..., n_8$ are values of cholera cases of all D's neighbours; and $d_4, d_5, ..., d_{11}$ are values of cholera cases of D. Indices of these variables indicate points in time. The model's first training data set is $\{w_1, w_2, w_3, n_1, n_2, n_3, d_4, d_5, d_6\}$ and first testing data set is $\{w_4, w_5, w_6, n_4, n_5, n_6\}$. The outcome of this model is $\{d_7, d_8, d_9\}$. Point in time to start training the CP model is 6.

5 Prediction Results Analysis

We built 29×3 RF regression models based on the FS data sets for 29 districts. Measures for regression models assessment are often root mean square error (RMSE) and adjusted coefficient of determination (adj-R^2) [20]. We calculate RMSE and adj-R^2 for all 29×3 models. In the following subsections, we first compare effects of weather and geographical predictors based on CP, WI and GI models and the two measures RMSE, adj-R^2. We then perform statistical analysis to find relationship of prediction accuracy and prediction length; and evaluate the importance of weather variables in 29×3 RF models.

5.1 Effects of Weather Predictors and Geographical Predictors to the Accuracy of Prediction

To compare the effect of weather and geographical predictors to prediction accuracy, measured in RMSE and adj-R^2, we use Tukey method for 3, 7, 14 and 30-day in advance prediction.

RMSE comparison in Fig. 6 does not show any statistical difference in models' prediction accuracy: all the 95 % confidence interval contains 0. In addition, p-values of RMSE comparison model are larger than 0.05. Thus using RMSE we cannot define which model among CP, WI and GI is the best. We will use adj-R^2 for models comparison.

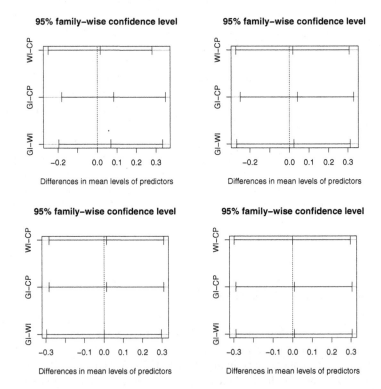

Fig. 6 Tukey multiple comparison of RMSE of CP, WI and GI models. From *left* to *right*, *top* to *bottom*: Comparison for 3, 7, 14 and 30-day in advance prediction

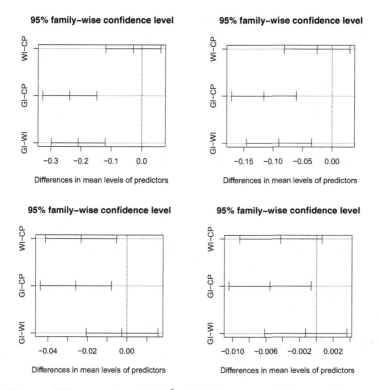

Fig. 7 Tukey multiple comparison of adj-R^2 of CP, WI and GI models. From *left* to *right*, *top* to *bottom*: Comparison for 3, 7, 14 and 30-day in advance prediction

With reference to GI-CP and WI-CP means and confidence interval in Fig. 7, we see that the CP models with highest adj-R^2 are the best. The GI models having the smallest adj-R^2 are the worst. It means that the number of cholera case in neighbours of a district strongly influences the number of cholera cases in that district. Figure 8 compares 3-day in advance prediction accuracy of CP, WI and GI models for Badinh, Dongda, Socson, Thanhxuan and Unghoa districts.

5.2 Relationship of Prediction Accuracy and Prediction Length

As mentioned above, the CPs are best models. We then apply CP models for final prediction of cholera cases on 29 districts of Hanoi, compare prediction results to observed data for $l = 3, 7, 14, 30$ and calculate adj-R^2 measures. Details of adj-R^2 measures are in Table 3. To observed change of prediction accuracy versus length of

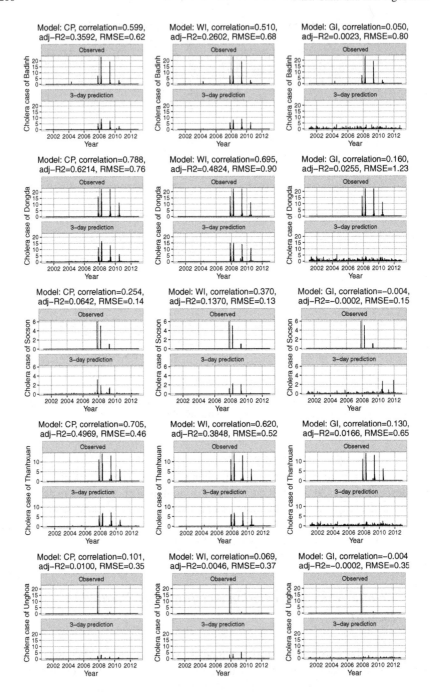

Fig. 8 Comparison of 3-day in advance prediction accuracy of CP, WI and GI models for Badinh, Dongda, Socson, Thanhxuan and Unghoa districts

Table 3 Adj-R^2 of each district for difference prediction length using CP models

District	Prediction length			
	30-day	14-day	7-day	3-day
Badinh	0.0092	0.0448	0.3126	0.3592
Bavi	−0.0001	0.0015	−0.0002	0.0186
Caugiay	0.0107	0.0623	0.2323	0.4142
Chuongmy	0.0015	0.1117	0.0745	0.2343
Danphuong	0.0003	0.0088	0.0177	0.2331
Donganh	−0.0002	0.0434	0.0085	0.0476
Dongda	0.0491	0.1325	0.4133	0.6214
Gialam	−0.0002	0.0007	0.0509	0.0831
Hadong	0.0106	0.0314	0.2248	0.3998
Haibatrung	0.0085	0.0719	0.209	0.4544
Hoaiduc	0.0037	0.0386	0.2188	0.5297
Hoankiem	0.024	0.0739	0.2466	0.4122
Hoangmai	0.0254	0.0458	0.1821	0.3372
Longbien	0.007	0.0121	0.1152	0.2316
Melinh	−0.0001	−0.0002	0.0666	0.0017
Myduc	−0.0002	0.0171	0.0169	0.3762
Phuxuyen	−0.0001	0.0003	0.0335	0.0281
Phuctho	0.0005	0.0214	0.0248	0.013
Quocoai	0.0012	0.0161	0.0553	0.0785
Socson	0.002	0.007	−0.0002	0.0642
Sontay	−0.0002	−0.0002	−0.0002	−0.0002
Tayho	0.0086	0.0735	0.2706	0.4905
Thachthat	0.0005	−0.0002	0.0051	0.15
Thanhoai	0.0094	0.0439	0.1588	0.2535
Thanhtri	0.0152	0.0159	0.0386	0.0967
Thanhxuan	0.0277	0.0551	0.287	0.4969
Thuongtin	0.0196	0.1813	0.1243	0.3211
Tuliem	0.0137	0.0699	0.2345	0.4197
Unghoa	−0.0002	0.0137	0.0008	0.01

prediction, we built a multiple linear regression model. The model's predictors are districts and the number of day to predict in advance. Outcome of the model is the adj-R^2 measure. Details of the model is in Listing 1.

Listing 1 Model for relationship of prediction accuracy and prediction length.

```
## Call:
## lm(formula = adjrsq ~ district + day, data = tmp)
##
## Residuals:
##       Min        1Q    Median        3Q       Max
## -0.16773  -0.06615  -0.01686   0.05959   0.25176
##
## Coefficients:
##                    Estimate  Std. Error  t value  Pr(>|t|)
## (Intercept)       0.2846737   0.0538533    5.286  9.34e-07  ***
## districtbavi     -0.1765346   0.0740029   -2.386    0.0193  *
## ...
## districtunghoa   -0.1753860   0.0740029   -2.370    0.0200  *
## day              -0.0076452   0.0009427   -8.110  3.17e-12  ***
## ---
## Signif. codes:  0 '***' 0.001 '**' 0.01 '*' 0.05 '.' 0.1 ' ' 1
##
## Residual standard error: 0.1047 on 86 degrees of freedom
## Multiple R-squared:  0.6179,  Adjusted R-squared:  0.4891
## F-statistic: 4.796 on 29 and 86 DF,  p-value: 7.651e-09

confint(fit, 'day')
##               2.5%        97.5%
## day   -0.009519223  -0.005771201
```

The multiple linear regression model shows that, if everything else remains and prediction length increases by one day, the adj-R^2 measure will decrease by 0.0076 with $[-0.0095, -0.0057]$ 95 % confidence interval. The p-value is very small (3.17^{-12}) indicating the statistical significant of the multiple linear model. The model's explained variation is 49 %. Figure 9 compare accuracy of CP models for 3, 7 and 14-day in advance prediction.

5.3 Importance of Weather Variables

From the 29 CP models built by RF regression method, we easily extract the importance of weather variables. The boxplots in Fig. 10 show that the daily mean temperature, relative humidity are the most important weather variables with approximately 50 % importance in comparison to other weather variables and cholera case variables. The precipitation and sun hour variables are about 30–35 % importance. The wind speed and SOI are less than 20 % importance.

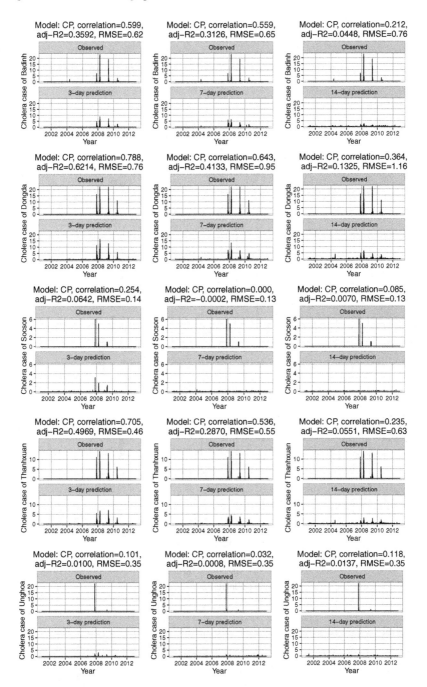

Fig. 9 Comparison of accuracy of CP models for 3, 7 and 14-day in advance prediction for Badinh, Dongda, Socson, Thanhxuan and Unghoa districts

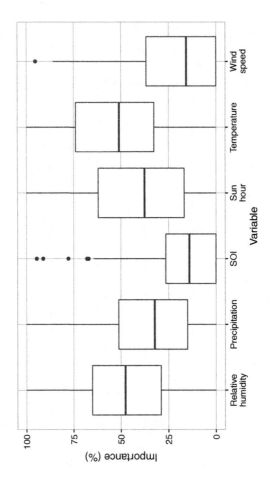

Fig. 10 Importance of weather variables to CP models

6   Conclusions

We have successfully built supervised learning models based on RF regression method for short-term prediction of cholera outbreaks in Hanoi from 2001 to 2012. To the best of our knowledge, this research is probably the first one that uses a machine learning approach to predict cholera outbreaks in Hanoi, Vietnam. The models built in this paper show that geographical factors of Hanoi, here are the numbers of cholera cases of a district of interest and its geographical neighbours, are very important ones to predict cholera outbreaks besides the weather variables. Among weather variables, daily mean temperature and daily mean relative humidity are the most important. SOI is least important weather factor to cholera in Hanoi. We also found that the prediction accuracy of our models, measured in adj-R^2, will decrease by 0.0076 if prediction length increases by one day.

Acknowledgments The authors wish to thank Assoc. Prof. Nguyen Ha Nam from VNU University of Engineering and Technology for his discussions and recommendations during experiments.

References

1. Ali, M., Lopez, A.L., You, Y.A., et al.: The global burden of cholera. Bull. World Health Organ. **90**(3), 209–218A (2012)
2. Sack, D.A., Sack, R.B., Nair, G.B., Siddique, A.K.: Cholera. Lancet **363**(9404), 223–233 (2004)
3. Nguyen, B.M., Lee, J.H., Cuong, N.T., et al.: Cholera outbreaks caused by an altered Vibrio cholerae O1 El Tor biotype strain producing classical cholera toxin B in Vietnam in 2007 to 2008. J. Clin. Microbiol. **47**(5), 1568–1571 (2009)
4. The World Health Organization: Cholera. Switzerland, Geneva (2003)
5. Kelly-Hope, L.A., Alonso, W.J., Thiem, V.D., et al.: Temporal trends and climatic factors associated with bacterial enteric diseases in Vietnam 1991–2001. Environ. Health Perspect. **116**(1), 7–12 (2008)
6. Emch, M., Feldacker, C., Islam, M.S., Ali, M.: Seasonality of cholera from 1974 to 2005: a review of global patterns. Int. J. Health Geograph. **2008**, 7–31 (2008)
7. Mendelsohn, J., Dawson, T.: Climate and cholera in KwaZulu-Natal. South Africa: the role of environmental factors and implications for epidemic preparednessi. Int. J. Hygiene Environ. Health **211**(1–2), 156–162 (2008)
8. Reyburn, R., Kim, D.R., Emch, M., Khatib, A., von Seidlein, L., Ali, M.: Climate variability and the outbreaks of cholera in Zanzibar. East Africa: a time series analysis. Am. J. Trop. Med. Hygiene **84**(6), 862–869 (2011)
9. Islam, M.S., Sharker, M.A., Rheman, S., et al.: Effects of local climate variability on transmission dynamics of cholera in Matlab, Bangladesh. Trans. R. Soc. Trop. Med. Hygiene **103**(11), 1165–1170 (2009)
10. Kovats, R.S., Bouma, M.J., Hajat, S., Worrall, E., Haines, A.: El Nino and health. Lancet **362**(9394), 1481–1489 (2003)
11. The World Health Organization: Using Climate to Predict Infectious Disease Outbreaks: A Review. Switzerland, Geneva (2004)
12. The World Health Organization, Control GTFoC: Cholera Country Profile: Vietnam. Geneva, Switzerland (2008)

13. The World Health Organization: Outbreak news: Severe acute watery diarrhoea with cases positive for Vibrio cholerae, Vietnam. Releve epidemiologique hebdomadaire/Section d'hygiene du Secretariat de la Societe des Nations = Weekly epidemiological record /Health Section of the Secretariat of the League of Nations, vol. 83, no. 18, 157–158 (2008)
14. Ali, M., Kim, D.R., Yunus, M., Emch, M.: Time series analysis of Cholera in Matlab, Bangladesh, during 1988–2001. J. Health Popul. Nutr. 31(1), 11–19 (2013)
15. Reiner, R.C., King, A.A., Emch, M., Yunus, M., Faruque, A.S.G., Pascual, M.: Highly localized sensitivity to climate forcing drives endemic cholera in a megacity. Proc. Natl. Acad. Sci. USA 109, 2033–2036 (2012)
16. Min, X., ChunXiang, C., DuoChun, W., Biao, K., Jia HuiCong, X., YunFei, L.X.W.: District prediction of cholera risk in China based on environmental factors. Chin. Sci. Bull. 58(23), 2798–2804 (2013)
17. Min, X., Cao, C., Wang, D., Kan, B.: Identifying environmental risk factors of Cholera in a coastal area with geospatial technologies. Int. J. Environ. Res. Public Health 12, 354–370 (2015)
18. Emch, M., Feldacker, C., Yunus, M., et al.: Local environmental predictors of Cholera in Bangladesh and Vietnam. Am. J. Trop. Med. Hygiene 78(5), 823–832 (2008)
19. Daily SOI data set of the Queensland, Australia. https://www.longpaddock.qld.gov.au/seasonalclimateoutlook/southernoscillationindex/soidatafiles/DailySOI1887-1989Base.txt
20. Hyndman, R., Athanasopoulos, G.: Forecasting: Principles and Practice. Otexts (2013)

Part V
Feature Extraction

Effect of the Text Size on Stylometry—Application on Arabic Religious Texts

S. Ouamour, S. Khennouf, S. Bourib, H. Hadjadj and H. Sayoud

Abstract In stylometry, there are two important technical questions: Firstly, does the text size affect the authorship attribution performances? and secondly, what could be the effect of the language on that attribution? To respond to those questions, we have conducted several experiments of authorship attribution applied on multi-size text documents. The text size varies from 100 words to 3000 words per document. For that purpose, a specific Arabic dataset has been conceived (i.e. A4P corpus). The corpus is made available for the scientific community and is suitable for the task of stylometry since the genre and theme are quite similar. Two types of features are investigated: character n-grams and words, in association with several classifiers, namely: SVM, MLP, Linear regression, Stamatatos distance and Manhattan distance. During the experiments, 2 types of scores are proposed: the "*Score of Good Attribution*" and "*Robustness against Size Reduction*" ratio. Results are quite interesting, showing that the minimum text size required for performing a fair authorship attribution, depends on the feature and classification method that are employed. For the evaluation task, a specific application of authorship attribution has been conducted on 7 religious books, where the main purpose was to check whether the Quran and Hadith could have the same Author or not. Results have clearly shown that those two books should have 2 different Authors.

S. Ouamour (✉) · S. Khennouf · S. Bourib · H. Hadjadj · H. Sayoud
Electronics & Computer Engineering Faculty, USTHB University,
Bab Ezzouar, Algeria
e-mail: siham.ouamour@uni.de; siham.ouamour@gmail.com

S. Khennouf
e-mail: khenouf.salah@gmail.com

S. Bourib
e-mail: bouribsamira@gmail.com

H. Hadjadj
e-mail: hadjadj.has@gmail.com

H. Sayoud
e-mail: halim.sayoud@uni.de

© Springer International Publishing Switzerland 2016   215
T.B. Nguyen et al. (eds.), *Advanced Computational Methods
for Knowledge Engineering*, Advances in Intelligent Systems
and Computing 453, DOI 10.1007/978-3-319-38884-7_16

Keywords Natural language processing · Authorship attribution · Stylometry · Performances versus size · Classifiers · Arabic language

1   Introduction and Related Work

Authorship Attribution (*AA*) is a research field concerned with the automatic classification of text documents with regards to their author(s). It tries to respond to the following question: Who is the author of this document?

Many studies have been reported during the last decades as described in [1–3], where many disputes were reported and several types of features and techniques were proposed too. In most cases the amount of data was big enough to bring significant characteristics to the author.

However, in 2001, de Vel et al. [4] reported a challenging research work of author identification on small texts such as in emails. The problem in such works is: was the small data provided by an email sufficient enough to make a fair author identification? Even though several works conducted on email documents reported quite good results too, their works were not so convincing since the scientific community agree that an efficient authorship attribution requires a quite big amount of text. In other words: the longer is the text, the higher is the precision of authorship attribution.

Now, what could be the minimum/optimum data size for that purpose? Some recent studies developed by Kim Luyckx and Walter Daelemans in 2011 [5] tried to investigate the size issue and showed the real effect of author data size in authorship attribution. Afterward, an interesting work was reported by Maciej Eder in 2013 [6], giving several responses to the problem by providing some key solutions to the minimum data size required for different cases and different languages, except for the oriental languages such as Arabic for instance (*i.e. those types of languages were not investigated*).

The results of Eder were interesting and useful, but we cannot extend them to all the languages that were not investigated, such as Arabic. Moreover, certain features were not tested by the author, which make his results, even though interesting, not extensible to every feature or classifier either.

That is, by the present investigation, we try to find out the minimum text size required to get a consistent authorship characterization for the Arabic Language. For that purpose, different types of features, distances and classifiers are tested and commented.

2   Dataset

The corpus used in our experiments consists of 45 text documents from 5 different authors, namely: 9 texts per author. These texts were written by five ancient Arabic philosophers (*without translation*) from various regions and date from the 11th, 12th and 13th centuries. Moreover, those texts have somewhat the same genre and topic. The original texts are quite long, but are divided herein into medium and short texts, so that we get several fixed sizes: the shortest text is about only 100 words and the longest one is about 3000 words. Hence, the different text sizes are 100, 500, 1000, 1500, 2000, 2500 and 3000 words per text.

Those different texts (*the 5 philosophers*) are extracted from the Universal Library (*Elwaraq*). During our experiments, the corpus was arbitrarily divided into two subsets, one subset for the training and another one for the testing. The training corpus, used to train each author model is composed of 10 long texts of 3000 words and the testing corpus is composed of 35 different texts.

To our knowledge, most researchers, in the literature, used their own text corpus for the task of author attribution. Although the Greek corpus of [7] and the Chinese corpus of [8] were included in Keselj works [9] for instance, unfortunately they were forced, like us, to assemble their Arabic corpus from classical texts (*eg. Ibn Roched, Elfarabi, etc.*). Finally, for purposes of further comparison, the entire corpus, which we built (*A4P*), has been made freely available on our personal website.

3   Authorship Attribution Methodology

In this section, we will explain the different steps of our approaches and the overall experimental protocol. As explained previously, two types of features (i.e. words and character 5-g) are used and five types of classifiers are employed (i.e. SVM, MLP, Linear Regression, Stamatatos distance and Manhattan distance).

3.1   General Algorithm

The general steps used for modeling the known authors (*training*) and identifying an unknown document (*testing*), during this investigation, are described in the following (*Pseudo-Code*) algorithm:

```
Setup: Begin
               Choose a Classifier: 1-SVM, 2-MLP, 3-LR, 4-Stam, 5-
Manh
       Setup: End

Features:      Begin
               Read Tex File
               Extract the Words
               Extract the Characters
               Concatenate Characters into Character-5Gram
Features:      End

Training:      Begin
       For every known Author (j=1...N):
               Do Features Extraction
               Build the Author Model
       End For
Training: End

Testing: Begin
       For every unknown author:
               Do Features Extraction
               Make an A.A. Classification
               Determine the most probable Author (Aj)
       End For
Testing: End

Result: Begin
       Display Result (Aj)
       Display the used Classifier
       Display the used Feature
Result:        End
```

As we can see in the previous algorithm, the first step consists in selecting the appropriate classifier to employ, then, in the second step, we proceed with features extraction from the text document. Note that in order to handle Arabic documents, we must use a UTF8 encoding.

The third step consists in building the Authors models (i.e. known authors). In the fourth step, which represents the testing, we redo the same tasks and then inject all the features to the classifier to make a decision. Finally, according to the obtained decision, we display the author identity and compute the performances scores.

3.2 Conventional Classifiers

The 3 conventional classifiers are described here below.

Manhattan Distance
This distance [5] is very reliable in text classification. The corresponding distance between two vectors X and Y is given by the following formula:

$$d_{X,Y} = \sum_{i=1}^{n} |X_i - Y_i| \tag{1}$$

where n is the length of the vector.

In this investigation, the different samples of the training are employed to build the centroid vector, which will be used, as reference, to compute the required distance with the previous formula (*also called KNN method*). Manhattan distance is simple to implement and very efficient for text classification.

Stamatatos distance

This distance represents a judicious way to measure the similarity between two texts. The distance between two vectors f and g is given by the following formula:

$$\sum_{i=1}^{n} [2(f_i - g_i)/(f_i + g_i)]^2 \tag{2}$$

where n is the length of the vector.

Multi-layer Perceptron (MLP)

The MLP (*Multi-Layer Perceptron*) is a classical neural network classifier that uses the errors of the output to train the neural network [10]. The MLP can use different back-propagation schemes to ensure the training of the classifier. It is trained by the different texts of the training set, whereas the remaining texts are used for the testing task. Usually the MLP is efficient in supervised classification, however in case of local minima; we usually can get some errors of classification.

Sequential Minimal Optimization based Support Vector Machine (SMO-SVM)

In machine learning, support vector machines (*SVMs*) are supervised learning models with associated learning algorithms that analyze data and recognize patterns, which are used for classification and regression analysis. The basic SVM takes a set of input data and predicts, for each given input, which of two possible classes forms the output, making it a non-probabilistic binary linear classifier. Given a set of training examples, each marked as belonging to one of two categories, a SVM training algorithm builds a model that assigns new examples into one category or the other. A SVM model is a representation of the examples as points in space, mapped so that the examples of the separate categories are divided by a clear gap that is as wide as possible. New examples are then mapped into that same space and predicted to belong to a category based on which side of the gap they fall on.

In addition to performing linear classification, SVMs can efficiently perform non-linear classification using what is called the kernel trick, implicitly mapping their inputs into high-dimensional feature spaces.

The SVM is a very accurate classifier that uses bad examples to form the boundaries of the different classes [11]. Concerning the Sequential Minimal Optimization (*SMO*) algorithm, it is used to speed up the training of the SVM [12].

Linear Regression
Linear Regression is the oldest and most widely used predictive model. The method of minimizing the sum of the squared errors to fit a straight line to a set of data points was published by Legendre in 1805 and by Gauss in 1809. Linear regression models are often fitted using the least squares approach, but they may also be fitted in other ways, such as by minimizing the "lack of fit" in some other norms (*as with least absolute deviations regression*), or by minimizing a penalized version of the least squares loss function as in ridge regression.

4   Authorship Attribution Experiments

Two types of features are employed in our authorship attribution experiments: once, using character n-grams [13] (5-g) and another time, using words.

In order to estimate the Authorship Attribution performances, we define the score of good Authorship Attribution (*AAS*), as follows:

$$AAS = \frac{Number\ of\ texts\ that\ are\ well\ attributed}{Total\ number\ of\ tested\ documents} \qquad (3)$$

4.1   Authorship Attribution Using Character Penta-Grams

We present the AA performances of the different classifiers with their tendency curves using character penta-grams as features in the following Figs. (1, 2, 3, 4 and 5).

As we can see the score of 100 % has been reached with the size of 3000 words for almost all the classifiers and distances, showing the pertinence of the character penta-gram as feature. Another interesting result concerns the minimum text size for which the AA score remains still high ($\cong 100$ %). For this point, we noticed that the different classifiers present different behaviors. For instance, the MLP appears to be very interesting since its performances are stable and remain unchanged until the size of 1500 words (*i.e. a score of 100 % is noted for documents of 3000, 2500, 2000 and 1500 words*). The SMO-SVM and Linear Regression seem less interesting since their performances are stable until the size of 2500 words. Less accurately but with a great stability, the Centroid Manhattan distance presents a constant AA score of 80 % from 3000 words down to 1000 words. This interesting result shows that this last distance could be recommended for short text documents, since a limited amount of data seems to be quite sufficient to fairly recognize an author.

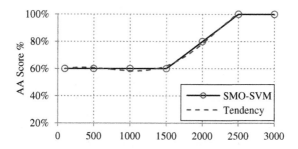

Fig. 1 Score of good authorship attribution with the SMO-SVM using character 5-g

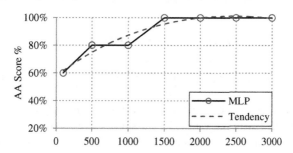

Fig. 2 Score of good author attribution with the MLP using character 5-g

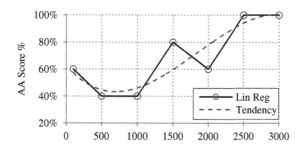

Fig. 3 Score of good authorship attribution of the linear regression using character 5-g

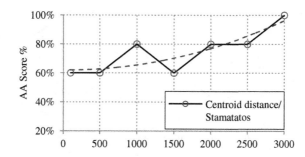

Fig. 4 Score of good authorship attribution of Stamatatos centroid distance using char. 5-g

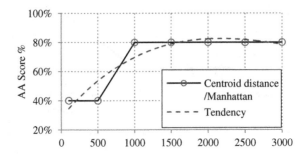

Fig. 5 Score of good authorship attribution of Manhattan centroid distance using char. 5-g

4.2 Authorship Attribution Using Words

We present the AA performances got by using the words as features (Figs. 6, 7, 8, 9 and 10).

In this second series of experiments, we can notice that quite good performances have been obtained with the Linear Regression and MLP (*score of 100 % for long texts*) showing that the size limit is about 2500 words per document. The SMO-SVM appears to be less accurate with a maximal AA score of 80 %.

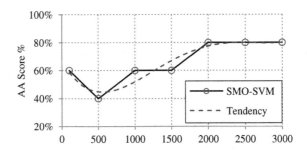

Fig. 6 Score of good authorship attribution of the SMO-SVM using words

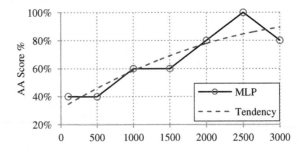

Fig. 7 Score of good authorship attribution of the MLP using words

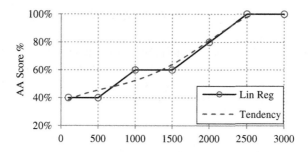

Fig. 8 Score of good authorship attribution of the linear regression using words

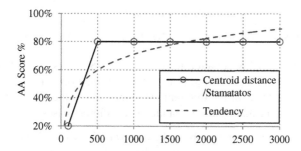

Fig. 9 Score of good authorship attribution of the Stamatatos centroid distance using words

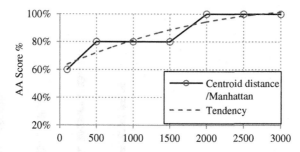

Fig. 10 Score of good authorship attribution of the Manhattan centroid distance using words

However, we notice that the best performances are provided by the Centroid Manhattan distance, for which the AA score remains equal to 100 % for 3000, 2500 and even 2500 words. Again, Stamatatos distance appears to be quite interesting too, due to its stability with regards to the size reduction. Hence a constant score of 80 % is reached for 3000, 2500, 2000, 1500, 1000 and even 500 words (*i.e. a constant score*).

4.3 Robustness Against Size Reduction Ratio: RSR

In order to assess the behavior of our attribution methods with regards to the size reduction, we defined a specific ratio, which we called "Robustness Against Size Reduction" ratio or RSR. The higher is this ratio, the more efficient is the attribution method with short texts. The RSR ratio is computed as follows.

$$RSR = \frac{\sum_{j=1}^{N} AAS_j.Size_j}{\sum_{j=1}^{N} Size_j} \qquad (4)$$

where N represents the number of documents (*with different sizes*) per author; and $Size_j$ represents the size of the jth text document. For instance, in our experiments, $Size_1 = 3000$, $Size_2 = 2500$, $Size_3 = 2000$, etc.... and then $\sum_{j=1}^{N} Size_j = 10600$.

By applying the RSR equation to the entire AAS scores, we get the following figure (Fig. 11). This figure shows interesting information, where we can visually evaluate the robustness of the different classifiers during the different experiments.

As we can see in this figure, we notice two different cases:

- For character 5-g, the most robust classifier (*i.e. the most suitable for short texts*) is the Multi-Layer Perceptron, with a RSR of 0.97 %;
- For words, it appears that Manhattan distance is the most robust classifier (*i.e. the most suitable for short texts*), with a RSR of 0.94 %.

Fig. 11 RSR ratio for the different classifiers and features

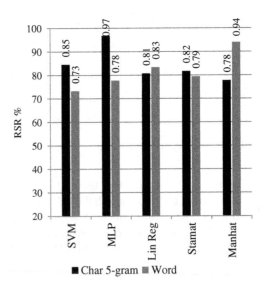

Consequently, for relatively short text documents (*i.e. less than 2000 words*), it would be highly recommended to employ those two combinations of classifier/ feature. In other words, when using character 5-g, the MLP appears to be the most interesting classifier to use in AA; and when using words, Manhattan distance appears to be the most interesting one. However, we limit this deduction to our set of features/ classifiers used in our experimental conditions.

5 Application of Authorship Attribution on Religious Books

In this application, seven Arabic religious books are investigated and analyzed in order to make a classification of the text documents per author: the experimented corpus is called *SAB-1*. Those texts have a size of about 2900 words per document so that the previous conditions (of Sect. 4) are well respected (>2500 words).

There are seven different books written by seven different authors: the holy Quran (book of God), Hadith (statements of the Prophet) and 5 other books written by 5 religious scholars, namely: *Mohammed al-Ghazali al-Saqqa, Yusuf al-Qaradawi, Omar Abdelkafy, Aaidh ibn Abdullah al-Qarni* and Amr Mohamed Helmi Khaled.

We also recall that several classifiers are employed in the experiments of authorship attribution by using the words as feature.

5.1 Experiments of Authorship Attribution Using Conventional Classifiers

In this section we report the different results obtained by using conventional classifiers. The different experimental results are exposed in Table 1.

Table 1 displays the different results obtained with the following classifiers: Manhatan centroid, MLP and SMO-SVM. Furthermore, the "Total identification error" summarizes the overall error of attribution for the 7 books. This indication gives us an interesting idea on the overall performances of authorship attribution.

We can see that the best classifier is the Manhattan centroid, which gives an error of only 1.05 %, the other classifiers present different performances (*total identification errors ranging between 1.05 and 4.2 %*). The three authors: Aaid-Alkarni, Abdelkafy and Alquaradawi present some problems of authorship attribution depending on the choice of the classifier. These two first ones are often confused with other authors. However, we can note that the Quran and Hadith books are attributed without any error (*error of 0 %*).

Table 1 Identification error in %

	Character-tetragram (%)	Word (%)
Manhatan centroid	1.05	1.05
MLP classifier	2.1	1.05
SMO-SVM classifier	3.1	2.1

5.2 Experiments of Authorship Attribution Using Fusion Techniques

In order to further enhance the authorship attribution performances, a fusion technique has been proposed and implemented, namely: the *Classifier-based Decision Fusion (CDF)*. The CDF technique fuses the decisions given by the different classifiers in order to produce a more accurate decision. Concerning the choice of the features, the word descriptor has been used because it has been shown that this type of feature presented relatively good performances during our experiments. Hence, We can see in Table 2 the corresponding results of this fusion technique.

As we can see in Table 2, by using the fusion technique, the authorship attribution error is equal to zero for every author. Te total identification score is 100 %, showing the superior performances of the fusion technique over the conventional classifiers as expected in theory. This result is very interesting since it shows that a combination of different features and/or classifiers can lead to high AA performances.

5.3 Comments

By observing the different experimental results, we can see that the 7 different books have been discriminated (*let us say*) correctly with regards to the writer/author: the corresponding text segments have been attributed to the correct authors with a small error of identification. Moreover, by using the fusion approach the attribution error have been reduced to 0 %. This important result shows that the classical features and classifiers that are usually employed in English and Greek languages got good results for the Arabic language too and appear to be utilizable for the authorship attribution of texts that are written in Arabic.

Table 2 Identification performances, in %, on the Arabic religious books

	Classifier-based decision fusion (%)	Number of books/authors
Identification error in %	0	7
Accuracy in %	100	7

The first conclusion we can state is that the fusion approach is quite interesting in multi-classifier or multi-feature authorship attribution.

Another important conclusion, one can deduce, is that the two religious books Quran and Hadith appear to have two different Authors.

6 Conclusion

In this investigation, we tried to respond to the following question: what could be the minimum text size for a given document to ensure a reliable AA task in Arabic language? To respond to that question, we conducted several experiments of authorship attribution applied on multi-size text documents: from 3000 words down to only 100 words per document. For that purpose, an Arabic dataset was constructed and collected from the books of 5 ancient Arabic philosophers.

Two types of features were investigated: character 5-g and words. Furthermore, several classifiers and distances were employed: SVM, MLP, Linear regression, Stamatatos distance and Manhattan distance.

Results have shown that the minimum text size required, for performing a fair authorship attribution, depends on the feature type and authentication method that are employed. But, in general, the size of 2500 words per document seems to be the minimum amount of textual data required for a fair AA in many cases. This result confirms the results reported by Eder in 2013 for the English language [6].

Furthermore, concerning the Robustness versus Size-Reduction, an interesting conclusion has been deduced. For instance, when using character 5-g, the MLP classifier appears to be one of the most robust classifier to use in AA; while when using words, Manhattan distance appears to be more suitable.

In the overall, character penta-grams seem to be more efficient than words, showing much better performances than this last one for almost all the classifiers used in this investigation. The unique exception is noticed with Manhattan distance, which presented a paradoxical result during the experiments.

Concerning the second application of stylometric analysis of the 7 religious books, results showed that the fusion technique has further improved the performances by reaching the accuracy of 100 % on the 7 books. By observing the obtained results, we could confirm that the 7 investigated books are stylistically distinct and should have different authors. As a consequence, we could also deduce that the Quran and Hadith should have 2 different Authors (i.e. The Quran's Author is different from the Hadith's Author).

Acknowledgements We warmly thank the research team of Dr. Juola and Al-Waraq library.

References

1. Juola, P.: JGAAP, Authorship attribution. In: Foundations and Trends in Information Retrieval, vol. 1, no. 3, pp. 233–334. Now Publisher (2006)
2. Stamatatos, E.: A survey of modern authorship attribution methods. J. Am. Soc. Inform. Sci. Technol. **60**(3), 538–556 (2009)
3. Sayoud, H.: A Visual analytics based investigation on the authorship of the holy Quran. In: 6th International Conference on Information Visualization Theory and Applications, pp. 177–181. Berlin, 11–14 Mar 2015
4. Vel, O. de., Anderson, A., Corney, M., Mohay, G.: ACM SIGMOD Rec. **30**(4), 55–64 (2001)
5. Luyckx, K., Daelemans, W.: The effect of author set size and data size in authorship attribution. Lit. Ling. Comput. **26**(1), 35–55 (2011)
6. Eder, M.: Does size matter? Authorship attribution, small samples, big problem. Lit. Ling. Comput. (2013). doi:10.1093/llc/fqt066
7. Stamatatos, E., Fakotakis, N., Kokkinakis, G.: Automatic authorship attribution. In: Proceedings of the 9th Conference of the European Chapter of the Association for Computer Linguistics, pp. 158–164 (1999)
8. Peng, F., Huang, X., Schuurmans, D., Wang, S.: Text classification in Asian languages without word segmentation. Proceedings of the sixth international workshop on Information retrieval with Asian languages **1**, 41–48 (2003)
9. Kešelj, V., Peng, F., Cercone, N., Thomas, C.: N-gram-based author profiles for authorship attribution. In: Proceedings of the Pacific Association for Computer Linguistics, vol. 3, pp. 255–264 (2003)
10. Sayoud, H.: Automatic speaker recognition—Connexionnist approach. PhD thesis, USTHB University, Algiers (2003)
11. Witten, I.H., Eibe, F., Trigg, L., Hall, M., Holmes, G., Cunningham S.J.: Weka: practical machine learning tools and techniques with Java implementations. In: Proceedings of the ICONIP/ANZIIS/ANNES'99 Workshop on Emerging Knowledge Engineering and Connectionist-Based Information Systems, New Zealand, pp. 192–196 (1999)
12. Keerthi, S.S., Shevade, S.K., Bhattacharyya, C., Murthy, K.R.K.: Improvements to platt's SMO algorithm for SVM classifier design. Neural Comput. **13**, 637–649 (2001)
13. Stamatatos, E.: On the robustness of authorship attribution based on character n-gram features. J. Law Policy **21**(2), 421–439 (2013)

Personalized Facets for Faceted Search Using Wikipedia Disambiguation and Social Network

Hong Son Nguyen, Hong Phuc Pham, Trong Hai Duong,
Thi Phuong Trang Nguyen and Huynh Minh Triet Le

Abstract The main aim of this paper is to deal with semantic search based on personalized facets using Wikipedia disambiguation data which can help to solve lexical ambiguity. User profile is learned from his/her activities and preferences in Facebook social network. Faceted graph visualization for result collaborative filtering is proposed. The facets are vertices representing ontological concepts. Other vertices represent instances belonging to the concepts, which are known as facets values. The vertices are highlighted by matching with user profile using TF-IDF feature vector model in order to individually produce search interfaces. The ties between vertices are ontological relations or properties considering as variables/attributes of facets. An algorithm to construct the faceted graph visualization and collaboratively filter search result is also provided. The faceted search method presented here is implemented to demonstrate these ideas.

Keywords Faceted search · Personalization · Wikipedia disambiguation · Semantic search · Social network

H.S. Nguyen (✉)
Faculty of Information Technology, Le Quy Don University, Hanoi, Vietnam
e-mail: son_nguyenhong2002@yahoo.com

H.P. Pham
Nguyen Tat Thanh University, Ho Chi Minh City, Vietnam
e-mail: phphuc7989@gmail.com

T.H. Duong · H.M.T. Le
International University, Vietnam National University, Ho Chi Minh City, Vietnam
e-mail: haiduongtrang@gmail.com

H.M.T. Le
e-mail: trietle95@gmail.com

T.P.T. Nguyen
Banking University of Ho Chi Minh City, Ho Chi Minh City, Vietnam
e-mail: phuongtrangict@gmail.com

© Springer International Publishing Switzerland 2016   229
T.B. Nguyen et al. (eds.), *Advanced Computational Methods
for Knowledge Engineering*, Advances in Intelligent Systems
and Computing 453, DOI 10.1007/978-3-319-38884-7_17

1 Introduction

The brain of human being connects the knowledge into a huge network of ideas, memories, definitions, perceptions. We can fully understand the meaning of a word even though it has ambiguous meaning, we can understand the same thing based on various terms. It is easy for human being but it is a big problem for computer. The major problem is shortage of specification of semantic heterogeneousness and ambiguity. We aim to look the "apple fruit" by putting the word "apple" in today famous search engines including Google, Yahoo! and Bing. We could not find the page which mentions "apple fruit" easily; most of the top results are about the "Apple Inc.", an American multinational corporation. All Google, Yahoo! or Bing are search engine not knowledge engine. They are good at returning a small number of relevant documents from a tremendous source of webpages on the Internet; but they still experience the lexical ambiguity issue, the presence of two or more possible meanings within a single word. Another problem with today search engines is that their filtering for their search results is not enough. In the process of finding the word "apple", we have observed that three big search engine only provide one visible criterion which is time to refine the result. In summary, there are two problems in today search engines:

- Lexical ambiguity: the question is that can the search engine return the correct meaning of the word that we are looking for when limited information of search query is provided?
- Search results filtering: Can the search engine offer better search results filtering such as collaborative filtering or content-based filtering?

In recent years, we have witnessed the emergence of Human-Computer information retrieval program where user can interact with the program to bring more complex information-seeking tasks.[1] Facets play the major part of this program. It is a way of classifying information and it also helps to solve the weakness of earlier knowledge representations. The faceted classification has been developed by scientists to offer an approach of knowledge representation which are rich and practical. However, the Faceted classification is only a solution to knowledge representation. We also need a mean which help to utilize that information, that mean is called Faceted Search. Faceted Search is becoming more and more popular especially in online shopping sites and site search. However, the facet types (category, price, brand, etc.) and the possible values for each facet are usually manually defined for a specific e-commerce site. For general purpose retrieval, automatic facet and facet-value recommendations are needed [1].

Facebook is now the most famous social networking site. It contains an extensive data of each member. The availability and extent of the profile data depends on the user's attitude towards entering and making the information visible in his profile [2]. The following data from Facebook can be utilized as user preferences: age, gender, group, geography data, posts, comments and likes. In Facebook, when user clicks the

---

[1]http://www.alexa.com/siteinfo/wikipedia.org.

"Like" button on an object such as fan page. That object is stored as an item in user profile. And certainly, the first thing that comes to mind when thinking about user preferences is likes as they explicitly express affinity. The same is for groups where many people with same interest gather as a small community. The geography data can point out the location-related affinity. In general, with the help of social networks, the many online services have the opportunity to know the users and get close to them in order to provide more relevant results. The valuable insights from networking sites are useful in many areas such as searching, recommendation or advertising.

Wikipedia is a free access and free content internet encyclopedia. Wikipedia is ranked as among the most popular website and constitutes the Internet's largest and most popular general reference work. In Wikipedia, internet users can freely create or edit a Wikipedia page's content. This "freedom of contribution" has a positive impact on both the quantity (fast-growing number of articles) and the quality (potential mistakes are quickly corrected within the collaborative environment) of this online resource [1].

To address the mentioned problems, this article's approach is to build a smart search that will utilize wiki disambiguation data and the information of social networks from user, the search is supposed to return the most relevant results with collaborative filtering. This approach includes the following:

- Provide the search results related to user's intentions based on his social network data. In particular, the faceted search program will solve the lexical ambiguity problem in current search engines.
- Based on the search results, provide a filtering that assists user to choose the correct result.

2 Related Works

Dynamic queries defined as interactive user control of visual query parameters that generate a rapid animated visual display of database search results [2–14]. The authors emphasize the interface with outstanding speed and interactivities. Ahlberg and Shneiderman built Film Finder to explore the movie database [1]. The graphical design contains many interface elements; parametric search is also included in a faceted information space. However, the results of Film Finder returning to users are not proactive and users are still able to select unsatisfactory combination. Later on, Shneiderman and his colleagues addressed the above problem on query previews. Query Previews are prevent wasted steps by eliminating zero-hit queries. That is mean the parametric search is replace by faceted navigation. In general, it helps user to have an overview over the selected documents. The mSpace project [7] described as an interaction design to support user-determined adaptable content and describe three techniques, which supports the interaction: preview cues, dimensional sorting and spatial context. Parallax was developed by David Huynh, its interface provides a "set-based browsing", that extends faceted search to shift views between related sets of entities. When user is browsing a set of results, Parallax provides the connections to related entities along with filter-base for current search result. Parallax is more like a semantic-web browser than a faceted search, because it supports more general ontology. Parallax made an important step that makes semantic web explore able, using many of the same techniques that have made faceted search successful [4]. In our previous work [15], we proposed an effective method to build a faceted search for unstructured documents that utilizes wiki disambiguation data to build the semantic search space; the search is to return the most relevant results with a collaborative filtering. The faceted search also can solve the lexical ambiguity problem in current search engines.

3 Personalized Facets for Faceted Search Using Wikipedia Disambiguation and Social Network

We present a proposed personalized facets methodology for faceted search using Wikipedia disambiguation and social network:

- Phase 1: Data Preparation: The data was not available for our experiment. In order to obtain the Wikipedia disambiguation data, we decided to download the Wikipedia dumps file which contain Wikipedia contents. And we will edit it to find a suitable data for our method.
- Phase 2: Prepare User Profile: We made user profile from user Facebook profile. It contains an extensive data of each member.
- Phase 3: Search visualization: we present a semantic search method using facets and user profile to automatically provide the most expected results in Wikipedia Disambiguation.

3.1 Phase 1: Data Preparation

The disambiguation pages [16] in Wikipedia are extracted from the main page file; each disambiguation entity contains list of all existing Wikipedia article of the give word; for example Java, an island, a programming language, an animal. Each meaning is categorized into facets/sub-facets.

In the Fig. 1, facets and sub-facets of Java disambiguation are:

- Facets: Places, Animal, Computing, Consumables, Fictional characters, Music, People, Transportation, Other uses.
- Sub-facets: Indonesia, United States, Other are sub-facets of facet Places.

Contents [hide]
1 Places
1.1 Indonesia
1.2 United States
1.3 Other
2 Animal
3 Computing
4 Consumables
5 Fictional characters
6 Music
7 People
8 Transportation
9 Other uses
10 See also

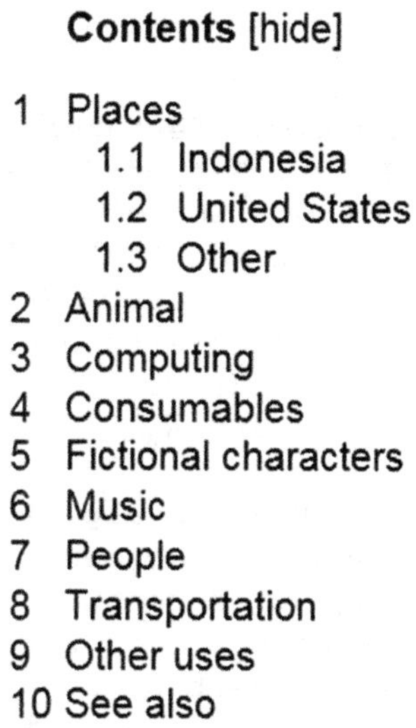

Fig. 1 Sample facet structure of java

In order to obtain the Wikipedia disambiguation data, we decided to download the Wikipedia dumps file which contain Wikipedia contents and load them into a MySQL database server (see Fig. 2).

After importing Wikipedia, a Disambiguation extraction process is carried out. All the disambiguation pages are extracted out from the database based on template of Disambiguation; for every disambiguation entity, an algorithm is applied to get the feature vector of its documents. Each document also includes its facet information. The process is as below Fig. 3.

MySQL database
Import
Wikipedia dumps
Disambiguation extraction

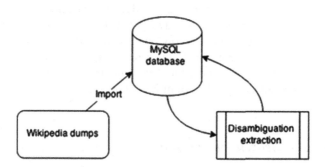

Fig. 2 Data preparation overview

Disambiguation
Set of documents
Set of document features

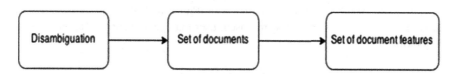

Fig. 3 Document features extraction process

The feature vectors of documents will also be used as search data.

```
Document features extraction algorithm
Input: Set of document, D
Output: Set of features, F
1. for each d ∈ D do
2.   Let f = GetFeaturealgorithm(d);
     /*converts document content into a feature vector */
3.   return f
4. return 𝓡w
```

```
Get Feature algorithm
Input: A document, d
Output: A feature vector
1. Remove stop words from d
2. Convert d into set of terms, T
4. for each t ∈ T do
   Remove verbs, get noun phrase
5. return feature vector contains list of noun phrases
```

3.2 Phase 2: Prepare User Profile

We use the information from Facebook page include: Category, Name, Description. In our method, only the Likes of user are used as user preferences which includes all the pages which have been liked by user [17, 18]. For each page, the associated information Name, Category and Description are extracted to form user profile data, the same will be used for matching with search results. We use Facebook Spring Social tool to enables the connection between our program with Facebook's Graph API which helps to get data from Facebook.

3.3 Phase 3: Search Visualization

The sample of partial Apple disambiguation

Figure 4 shows an example of a surfing user browses web to get more information about Apple Inc. which produces the iPhone and iPad that he is using. As modern search engines are keyword-based, user may get the correct "apple" that he is looking for.

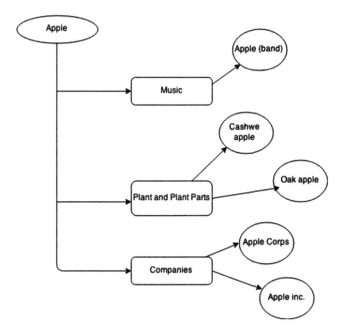

Fig. 4 Sample of partial Apple disambiguation

The search program is designed to solve the above user query. As query is entered, the program will return all the related result, the most relevant results are highlighted. Along with search result, the facet function is provided for result filtering, on selecting a facet, only its facet values are left.

Beside the search input, the user search graph interface includes two sections. The Facet graph which has a list of facets, the same is displayed in tree format gives user an overview about the result. Each facet has a list of available values associated with it. Every facet is displayed as a rectangle.

The second section Document Graph visualization is a graph to present results $\mathcal{R}(V_r, E_r)$ on which each vertex $v \in V_r$ presented with different sizes. Here, we proposed the Algorithm 1 [10] to compute the size of vertices. Let $\mathcal{R}_w\left(V_r', E_r'\right)$ be a new result after matching $\mathcal{R}$ with User Profile. Out of which, $V_r' = \{(v_i, w_i, s_i) | v_i \in V_r\}$ with s_i is the size and w_i is the weight of vertex v_i. This algorithm uses two loops to compute the weight of each vertex w_i and determine the maximum weight w_{max}. Then, it computes the size of each vertex (s_i) by using following equation:

$$s_i = S_{min} + w_i \times \frac{S_{max} - S_{min}}{w_{max}} \tag{1}$$

where $S_{max} S_{min}$ stand for the constant of minimum and maximum size.

Computing the sizes algorithm
Input: $\mathcal{R}(V_r, E_r)$
Output: $\mathcal{R}_w(V'_r, E'_r)$
Initialize $\mathcal{R}_w(V'_r, E'_r)$ with $V'_r = \emptyset$ and $E'_r = E_r$
Initialize $w_{max} = 0$
1.   for each$v \in V_r$do
2.   Let $v' = (v, w = 0, s = 0)$
3.   Let $weight = tfidf(v, P)$; // P is user profile feature
vector
         /*tfidf compute the $weight\, v$ on P */
4.   if $weight > w_{max}$ then
5. $w_{max} = weight$
6.   Adding v' into V'_r
7. $step = (S_{max} - S_{min})/w_{max}$
8.   for each$v' \in V'_r$do
9. $v'.s = S_{min} + v'.w * step$
10. return $\mathcal{R}_w$

4   Experiment

In this section we present experimental evaluations for our techniques. We first evaluate our disambiguation extraction process that we used to create data for our search program. Then, we evaluate the search results in term of efficiency, effectiveness and performance. Finally, we examine the algorithm which was used to compute the size of vertices. In order to implement the search program for the demonstration, we use Spring Tool Suite[2] as development tool.

4.1   Disambiguation Extraction Evaluation

There are 158613 disambiguation pages in total as of 11/2014. Each disambiguation page contains zero or more facets. The evaluation was carried out by comparing the raw data with extracted result. Two important aspects are considered during this duration: the first aspect is the number of extracted facets along with their structures in a one disambiguation entity so that our semantic search can benefit from this information, a list of facets is provided so that user can base on that to seek for correct documents; another thing is number of documents in one disambiguation

---

[2]https://spring.io/tools.

Table 1 Sample extracted data for places facet in Java disambiguation

Facet	Document
Java->Places->Indonesia	Java sea
Java->Places->Indonesia	Java trench
Java->Places->Other	Java (town)
Java->Places->Other	Java road
Java->Places->Other	Java, São Tomé and Príncipe
Java->Places->Other	Java district
Java->Places->Other	Java eiland
Java->Places->United States	Java, New York
Java->Places->United States	Java, Virginia
Java->Places->United States	Java, Ohio
Java->Places->United States	Java, South Dakota
Java->Places->United States	Coffee County, Alabama

entity, the information from these document are used for matching the query entered by user, the same are indexed to be used as the search data for the program. Below is a same extracted data for Places facet in Java (Table 1).

In Wikitext, raw content of Wikis page stored in database, the Facets are within double equal symbols "== ==", sub-facets are within triple equal symbols "=== ===", the documents start with asterisk symbols "*" and are within pairs of double square "[[]]" brackets. We based on those notations to pull out the data. Six sample disambiguation entities are taken for extraction "apple", "obama", "java", "joker", "iphone", "alien".

According to Table 2, the number of extracted facets are always the same as original one, however when extracting the documents from disambiguation page, the results are not completely correct, this is because the complexity of structure of page in raw format. In general, the result of extraction is quite accurate; it can retain the information from original disambiguation entity.

Table 2 Extracted facets/documents compared to original data

Disambiguation page	Extracted facets/original facets	Facets accuracy (%)	Extracted documents/original documents (%)	Documents accuracy (%)
Java	12/12	100	45/48	93.76
Apple	8/8	100	46/46	100
Obama	3/3	100	14/14	100
Joker	11/11	100	58/59	98.3
Iphone	0/0	100	14/14	100
Alien	6/6	100	39/39	100

4.2   Search Result Evaluation

In this section, we will evaluate our search program, we analyze the personalization aspect of the program, how it accommodates the differences between individuals. The accuracy is evaluated using precision and recall technique; finally we evaluate the satisfaction of user about our program.

4.2.1   Personalization Evaluation

To perform this evaluation, we created three test users; Facebook offers test user feature, a test user can experience the app as regular user but it is invisible to normal users, furthermore the app can be granted any permission from test user without the approval from Facebook. Our three test users have liked various pages in Facebook; Richard has interests in Java programming language, Apple Inc., and United States; for Tom, they are Indonesia, Java sea and Apple fruit. The last user, Patricia, has showed no interests on Facebook. In order to examize how the program reponses to different profiles, we use two queries: "java" and "apple" (Table 3).

Six tests were executed and the search program behaves differently for different profiles; in the search results, sizes of vertices for specific documents are varied for various users (Tom, Richard and Patricia).

4.2.2   Accuracy Evaluation

In the Fig. 5, the effectiveness of the returned results of our program is evaluated. In most of cases, all the relevant documents are retrieved but the accuracy is not very good, many times the irrelevant documents are shown to user.

4.2.3   Filtering Evaluation

Because of the problem when matching words together, there will be the cases that user is not able to get his interested results (Table 4); in these situation, user can refine search results by selecting facet or sub facets in the facet graph.

Table 3 Test users with different interests

User	Interests (Likes)
Richard	Java programming language, Apple Inc., United States
Tom	Indonesia, Java sea, Apple fruit
Patricia	None

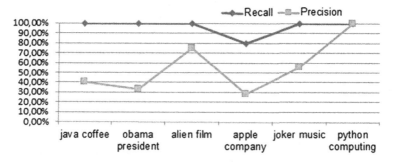

Fig. 5 Measure search result using precision and recall

Table 4 Search result summary

User	Query	Most highlighted vertices
Richard	Java	Java programming language, java software platform, java virtual machine
	Apple	Apple Inc., Apple II series, Apple store
Tom	Java	Coffee, java (an island), java (DC Comics)
	Apple	Apple (fruit), tomato, Apple Inc.
Patricia	Java	None (all vertices are at default size)
	Apple	None (all vertices are at default size)

In a common flow, user will be able to select the correct information after two steps in our search program; the first is to enter search query, next step is to refine the search result based on facet value on facet graph. This is considered very fast.

4.3 Vertices Size Effectiveness

To evaluate the algorithm that was used to compute the size of vertices, we compare the original graph (G1) and the matched graph (G2). In G1, let the size of each vertex be 75 px. And in G2, the size of each vertex is computed by using Eq. 1 with s_min = 50 px and s_max = 100 px.

The comparison was considered with tasks named apple. The detailed results of this comparison were presented in Fig. 1. To have more conviction, we expanded 10 other tasks. For each task, we compute the rate between the total size of G2 and G1:

$$r = \frac{\text{Total size of graph 2}}{\text{Total size of graph 1}} \times 100\%$$

A graph of rates is shown in Fig. 6.

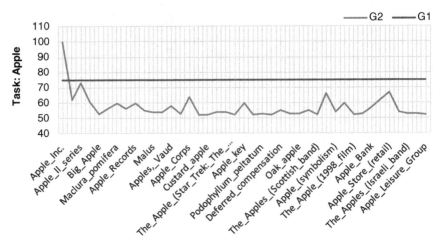

Fig. 6 Compare the size of G1 and G2

According to this results, we realize that G2 not only uses less resources (occupies less pixel on screen) than G1 does, but also facilitates exploring of the expected results.

5 Conclusions

We have presented a methodology regarding personalized facets for semantic search using two data sources, the first one is Wikipedia Disambiguation, a dataset that helps to solve lexical ambiguity; the other one is Facebook user profile. A complex process was carried out to make the data available for experiment from raw Wikis data. The graph with modern user interface was implemented to show individualized search results by matching facet values as documents with Facebook user profile.

In experiment, firstly we evaluate the effectiveness of our process to extract the disambiguation data from Wiki. The result shows that the process retains most of the information of disambiguation page from Wikipedia. Secondly the evaluations for effectiveness and efficiency our search program is conducted. The program not only helps to solve the ambiguity of words when searching but also help user to seek for the information quickly with few interactions.

References

1. Ahlberg, C., Shneiderman, B.: Visual information seeking: tight coupling of dynamic query filters with starfield displays. In: Adelson, B., Dumais, S., Olson, J. (eds.) Proceedings of the SIGCHI Conference on Human Factors in Computing Systems (CHI'94), pp. 313–317. ACM, New York, NY, USA (1994)
2. Brunk, S., Heim, P.: tFacet: hierarchical faceted exploration of semantic data using well-known interaction concepts. In: Proceedings of DCI 2011. CEUR-WS.org, vol. 817, pp. 31–36 (2011)
3. Heim, P., Ertl, T., Ziegler, J.: Facet graphs: complex semantic querying made easy. In: Proceedings of the 7th Extended Semantic Web Conference (ESWC 2010). LNCS, vol. 6088, pp. 288–302. Springer, Berlin, Heidelberg (2010)
4. Heim, P., Ziegler, J., Lohmann, S.: gFacet: a browser for the web of data. In: Proceedings of the International Workshop on Interacting with Multimedia Content in the Social Semantic Web (IMC-SSW 2008). CEUR-WS, vol. 417, pp. 49–58 (2008)
5. Heim, P., Ziegler, J.: Faceted visual exploration of semantic data. In: Human Aspects of Visualization. Lecture Notes in Computer Science, vol. 6431, pp. 58–75. Springer, Berlin, Heidelberg (2011)
6. Koren, J., Zhang, Y., Liu, X.: Personalized interactive faceted search. In: Proceedings of the 17th International Conference on World Wide Web, pp. 477–486. ACM New York, NY, USA (2008)
7. Schraefel, M.C., Karam, M., Zhao, S.: mSpace: interaction design for user-determined, adaptable domain exploration in hypermedia. In: AH 2003: Workshop on Adaptive Hypermedia and Adaptive Web Based Systems, pp. 217–235. Nottingham, UK (2003)
8. Thi, A.D.H., Nguyen, T.B.: A semantic approach towards CWM-based ETL processes. In: Proceedings of I-SEMANTICS, vol. 8, pp. 58–66 (2008)
9. Tunkelang, D.: Faceted Search. Morgan & Claypool Publishers (2009)
10. Wagner, A., Ladwig, G., Tran, T.: Browsing-oriented semantic faceted search. In: DEXA'11 Proceedings of the 22nd International Conference on Database and Expert Systems Applications, vol. 1, pp. 303–319. Springer, Heidelberg (2011)
11. Hostetter, C.: Faceted searching with apache solar. In: ApacheCon, US (2006)
12. Nguyen, T.B., Schoepp, W., Wagner, F.: GAINS-BI: business intelligent approach for greenhouse gas and air pollution interactions and synergies information system. In: iiWAS 2008, pp. 332–338 (2008)
13. Nguyen, T.B., Wagner, F., Schoepp, W.: Cloud intelligent services for calculating emissions and costs of air pollutants and greenhouse gases. ACIIDS **2011**, 159–168 (2011)
14. Le, T.,Vo, B., Duong, T.H.: Personalized facets for semantic search using linked open data with social networks. In: IBICA 2012, pp. 312–317 (2012)
15. Dang, B.D., Nguyen, H.S., Nguyen, T.B., Duong, T.H.: A framework of faceted search for unstructured documents using wiki disambiguation. In: ICCCI, vol. 2, pp. 502–511 (2015)
16. Mihalcea, R.: Using wikipedia for automatic word sense disambiguation. In: Proceedings of NAACL HLT, pp. 196–203 (2007)
17. Duong, T.H., Uddin, M.N., Li, D., Jo, G.S.: A collaborative ontology-based user profiles system. In: Proceedings of ICCCI'09, Social Networks and Multi-agent Systems, pp. 540–552. Springer, Heidelberg (2009)
18. Duong, T.H., Mohammed, N.U., Nguyen, D.C.: Personalized semantic search using ODP: a study case in academic domain. ICCSA **2013**, 607–619 (2013)

Readiness Measurement Model (RMM): Mathematical-Based Evaluation Technique for the Quantification of Knowledge Acquisition, Individual Understanding, and Interface Acceptance Dimensions of Software Applications on Handheld Devices

Amalina F.A. Fadzlah

Abstract This paper presents a mathematical-based evaluation technique as a new method in assessing the readiness of handheld application usage. This research considers specifically the quantification of readiness parameters useful to express and estimate the overall readiness of handheld application usage. As a result, a new and simple mathematical-based evaluation model for assessing the readiness of handheld application usage, namely Readiness Measurement Model (RMM), was established. The proposed model integrates three dimensions for evaluating handheld application usage readiness including knowledge acquisition, individual understanding, and interface acceptance.

Keywords Readiness evaluation · Assessment metric · Readiness measurement model · Handheld application

1 Introduction

The handheld device (also known as handheld computer or simply handheld), has been a growth area in computing for over ten years and its development continues today. These devices have become accessible and feature-rich for use in different fields, such as engineering and construction. The devices can be used for multiple

A.F.A. Fadzlah (✉)
Department of Computer Science, Faculty of Defence Science and Technology, Universiti Pertahanan Nasional Malaysia, Kem Perdana Sg Besi, Kuala Lumpur, Malaysia
e-mail: amalina.farhi@upnm.edu.my

© Springer International Publishing Switzerland 2016
T.B. Nguyen et al. (eds.), *Advanced Computational Methods for Knowledge Engineering*, Advances in Intelligent Systems and Computing 453, DOI 10.1007/978-3-319-38884-7_18

purposes. People of diverse academic and professional backgrounds can use these devices to run many types of application software, known as apps. Handheld applications were originally offered for general productivity and information retrieval. However, public demand and the availability of developer tools drove rapid expansion into other categories. The explosion in number and variety thus made handheld application usage a challenge, which in turn led to the creation of a wide range of readiness evaluation methods [1]. Past studies offer a variety of evaluations on the uses of handheld applications [2–14].

With the advancement of technology, there has been a convergence of computation and communication capabilities in mobile devices. There is a substantial growth of different applications. In case of handheld applications, battery life plays a crucial role. Hence, it is important for the designers to make architecture enhancements power efficient [11]. As per the fact, mobile devices have drastically boosted the enhancements in algorithms, being them a requirement of integration in various mobile applications, and for multi-purposes [15]. Although there are many methods to guide towards evaluation of handheld application usage readiness, unfortunately there is no effort to develop an evaluation model based on mathematical approaches. Mathematical-based modelling is an important and fundamental method for assessing readiness: not only to optimise the process and predict the results, but also to evaluate the burden of using handheld applications.

To circumvent the high demands of mathematical model-based evaluation methods, a reliable and accurate relative quantification model is needed in assessing the handheld application usage readiness. Therefore this research focuses on specification and measurement of readiness in order to derive a mathematically based scheme for evaluating handheld application usage outcomes. The advancement in technology has made it possible for the mobile devices to perform significant computations based on events triggered from sources of information and input devices. The events carry the information about the conditions and context of the user [14]. It facilitates the network nodes and the mobile devices to adapt the mode of communication on the basis of these events. This research considers the relative quantification of readiness parameters that usefully express and estimate quantitatively the overall readiness of handheld application usage. As a result, a new and simple mathematical-based evaluation model for assessing the readiness of handheld application usage, namely Readiness Measurement Model (RMM), was established.

2 Methods and Materials

The advancement in technology has made it possible for the mobile devices to perform significant computations based on events triggered from sources of information and input devices. The events carry the information about the conditions and context of the user [14]. It facilitates the network nodes and the mobile devices to adapt the mode of communication on the basis of these events. Ziarati et al. [9] proposed characteristics of virtual organization based on three axes. These included core competencies, cooperation instrument, and cooperation culture. These axes formed the basis for the assessment of the readiness of the company. The idea of virtual maturity indicates the preparedness of an organization to enter the world of virtual collaboration.

The advancement in technology has made it possible for the mobile devices to perform significant computations based on events triggered from sources of information and input devices. The events carry the information about the conditions and context of the user [14]. It facilitates the network nodes and the mobile devices to adapt the mode of communication on the basis of these events. Ziarati et al. [9] proposed characteristics of virtual organization based on three axes. These included core competencies, cooperation instrument, and cooperation culture. These axes formed the basis for the assessment of the readiness of the company. The idea of virtual maturity indicates the preparedness of an organization to enter the world of virtual collaboration.

Said et al. [10] conducted a study on the user technology readiness measurement. They analysed the fingerprint system for monitoring the attendance in the organisation. Statistical analysis was performed to study the variables optimism, innovativeness, insecurity, and discomfort. The findings of the study showed that user readiness is positively correlated with optimism and innovativeness. The findings also showed that insecurity and discomfort are negatively correlated with user readiness. The main purpose of this study was to develop a model describing a mathematical-based evaluation technique for assessing the readiness of using handheld applications. This model extended the conceptual and empirical relationship-driven Readiness Measurement Framework (RMF) developed by Fadzlah [8].

Figure 1 shows the structure of RMF, which describes users' needs and requirements when encountering the assessment of handheld application usage readiness. This framework discusses the three distinct attributes of Knowledge Acquisition, Individual Interpretation and Interface Acceptance. The first attribute, Knowledge Acquisition, enables users to effectively acquire the knowledge necessary to perceive readiness towards handheld application usage. This attribute is measured by the metrics Skills Integrated, Problems Solved, Meanings Understood, and Reasons Explained. The second attribute enables users to effectively interpret their individual perceptions of readiness towards handheld application usage. This Individual Interpretation attribute is measured by the metrics Answers Supplied, Examples Provided and Questions Given. The third attribute enables users to effectively accept the design of a user interface in perceiving readiness towards handheld application usage. This Interface Acceptance attribute is measured by the metrics Information Memorized, Representations Recalled, and Layouts Recognized.

A positive metrics value results in a positive attribute value, which ultimately leads to a positive readiness value in using handheld applications. Table 1 displays readiness attributes and standardised definitions of their contributing metrics. All attributes and metrics in the framework were found statistically significant in

Fig. 1 Readiness
Measurement Framework
(RMF)

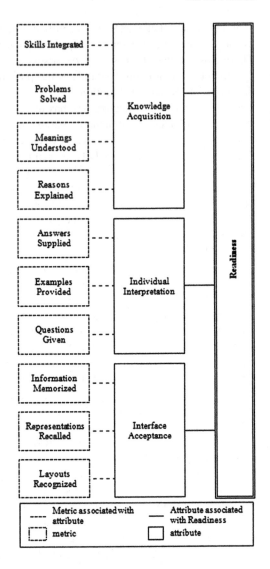

assessing the readiness of handheld application usage. In order to develop a
mathematical model in assessing the readiness of using handheld applications, this
study was designed to achieve five major objectives: develop scale items for
evaluating the importance of measuring the readiness of handheld application
usage, analyse the weightage between metrics and attributes towards assessing the
readiness of handheld application usage, determine suitable metrics and attributes
code with which to assess the readiness of handheld application usage, construct
equations of readiness metrics and attributes in order to assess the handheld
application usage, and finally develop a model within which to structure the
readiness measures in order to assess the handheld application usage readiness.

Table 1 Attributes and metrics of handheld application usage readiness

Attribute	Metric
Knowledge Acquisition	Skills Integrated depends on the number of skills integrated correctly
	Problems Solved depends on the number of problems solved correctly
	Meanings Understood depends on the number of meanings understood correctly
	Reasons Explained depends on the number of reasons explained correctly
Individual Interpretation	Answers Supplied depends on the number of answers supplied correctly
	Examples Provided depends on the number of examples provided correctly
	Questions Given depends on the number of questions given correctly
Interface Acceptance	Information Memorized depends on the amount of information memorized correctly
	Representations Recalled depends on the number of representations recalled correctly
	Layouts Recognized depends on the number of layouts recognized correctly

Technology gadgets, such as handheld devices, have proved quite efficient in the field work [12]. Examples include participant engagement and survey data collection. Such devices permit quick information sharing and evaluation with a low cost. The benefits of these devices can be acquired in improved data validation, reduced data collection errors, and reduced data cleansing time. Roth and Unger [13] proposed a platform that was developed for running groupware applications on handhelds. It was a novel idea because the usual approach in groupware applications is to assume desktop computer as the hardware platform. The findings showed significant differences in the execution of groupware applications due to the platform change. These included resource constraints, network stability, and privacy considerations.

2.1 Scale Development

The iterative development of Readiness Measurement Scale (RMS) was designed based upon a number of proposed metrics for measuring readiness of handheld applications. These metrics were collected and gathered by considering multiple theories to integrate both objective and subjective measures for readiness evaluation. The construction of RMS items further includes conceptual mapping based upon the principle that readiness can be measured by items of knowledge acquisition, individual understanding and interface acceptance dimensions [8].

The original items were modified to address the requirements for assessing the importance of measuring the readiness of handheld-based application system and

Table 2 Readiness
measurement items

Measurement item
Knowledge Acquisition
(I think, it is important to measure the … made by handheld application user within session or treatment)
… number of skills integrated correctly
… number of problems solved correctly
… number of meanings understood correctly
… number of reasons explained correctly
Individual Interpretation
(I think, it is important to measure the … made by handheld application user within session or treatment)
… number of answers supplied correctly
… number of examples provided correctly
… number of questions given correctly
Interface Acceptance
(I think, it is important to measure the … made by handheld application user within session or treatment)
… amount of information memorized correctly
… number of representations recalled correctly
… number of layouts recognized correctly

specific user tasks. For example, to modify the readiness metric into question, 'number of skills integrated correctly'. Thus result, 'I think, it is important to measure the number of skills integrated correctly by handheld application user within session or treatment' question. This RMS consisted of 10 items rated on a five-point Likert scale from extremely agree, slightly agree, neutral, slightly disagree and extremely disagree (refer Table 2).

2.2 Weightage Extraction

Weightage extraction was performed based on previous work regarding the identification and determination of measures for assessing the readiness of handheld application usage. This RMS was used to gather information from respective users to indicate their level of agreement towards the importance of each readiness measure, based on their experience and perception of handheld application usage. Data collected were entered into the statistical software program for analysis. Relationship evaluation tests carried out in the software program to determine the strength between measures in different hierarchical levels in order to assess the overall readiness of handheld application usage.

Data analyses were conducted using latest version of SPSS V.23 for descriptive analysis and relationship test. Result of the relationship test thus generalized weightage values for both metrics and attributes for measuring readiness of handheld application usage. Results for this study are presented in the following order of

Table 3 Handheld application usage background

Usage background		n	%
Expertise	Beginner	6	1.6
	Intermediate	126	32.8
	Advanced	141	36.7
	Expert	111	28.9
Experience	Less than 1 year	21	5.5
	Between 1 and 3 years	45	11.7
	Between 3 and 5 years	78	20.3
	More than 5 years	240	62.5
Duration	Less than 1 h	24	6.3
	Between 1 and 3 h	40	10.4
	Between 3 and 5 h	107	27.9
	More than 5 h	213	55.5
Frequency	Rarely	30	7.8
	Often	81	21.1
	Sometimes	84	21.9
	Always	189	49.2

descriptive analysis, confirmation of premier weightage, linearity code of parameters, equation of readiness metrics and attributes as well as mathematical model for readiness quantification.

A total number of 397 targeted participants responded. After exclusion of duplicate entries and missing entries (more than 3.27 % of incomplete data), there were 384 valid responses. This study used list wise deletion for missing and duplicate data, therefore only valid responses were used. The perceived mobile usage competency of the respondents was high. Results reported more than 50 % of respondents somewhat agreeing, strongly agreeing and extremely agreeing that they were competent. The handheld application usage backgrounds of the respondents are summarized in Table 3.

The relationship tests revealed that the Skills Integrated, Problems Solved, and Meanings Understood metrics were all weighted towards the Knowledge Acquisition attribute with a value higher than 0.5 at 0.580, 0.527, and 0.564 respectively, with only the Reasons Explained (0.463) metric being below 0.5. As the results of the relationship strength tests revealed, Skills Integrated, Problems Solved, Meanings Understood and Reasons Explained metrics contributed sequentially towards the Knowledge Acquisition (KA) attribute in assessing the readiness of handheld application usage, where weight values for subsequent metrics were symbolised as w_{KA1}, w_{KA2}, w_{KA3}, and w_{KA4}. The Answers Supplied (0.553), Examples Provided (0.579) and Questions Given (0.622) metrics were similarly weighted higher than 0.5 which was interpreted as having a strong relationship value towards the Individual Interpretation attribute in assessing the handheld application usage readiness. These metrics, Answers Supplies ($m = 1$), Examples Provided ($m = 2$), and Questions Given ($m = 3$), which were found to contribute

Table 4 Weight value and symbol of readiness metrics

Attribute	Metric	Value	Symbol
Knowledge Acquisition	Skills Integrated *(number of skills integrated correctly)*	0.580	w_{KA1}
	Problems Solved *(number of problems solved correctly)*	0.527	w_{KA2}
	Meanings Understood *(number of meanings understood correctly)*	0.564	w_{KA3}
	Reasons Explained *(number of reasons explained correctly)*	0.463	w_{KA4}
Individual Interpretation	Answers Supplied *(number of answers supplied correctly)*	0.553	w_{II1}
	Examples Provided *(number of examples provided correctly)*	0.579	w_{II2}
	Questions Given *(number of questions given correctly)*	0.622	w_{II3}
Interface Acceptance	Information Memorized *(amount of information memorized correctly)*	0.388	w_{IA1}
	Representations Recalled *(number of representations recalled correctly)*	0.442	w_{IA2}
	Layouts Recognized *(number of layouts recognized correctly)*	0.309	w_{IA3}

significantly towards the Individual Interpretation (*II*) readiness attribute, were labelled w_{II1}, w_{II2}, and w_{II3} respectively to represent the weight value of metrics which contributed to attribute Individual Interpretation.

In contrast, the Information Memorized (0.388), Representations Recalled (0.442), and Layouts Recognized (0.309) metrics were found to contribute towards the Interface Acceptance attribute with a weight of less than 0.5, which means that these metrics were having less strength with smaller extent in influencing the readiness of handheld application usage. Finally, according to the obtained relationship weight results, sequentially labelled as w_{IA1}, w_{IA2}, and w_{IA3}, Information Memorized, Representations Recalled, and Layouts Recognized were found first, second and third metrics respectively to contribute towards the Interface Acceptance (*IA*) attribute value in assessing the readiness of handheld application usage. In summary, only six readiness metrics were identified as having a high relationship weight value (i.e., over 0.5) towards their corresponding readiness attributes, with the remaining four readiness metrics associated with a weight value lower than 0.5. The extraction of weight value as well as weight symbol for each readiness metric towards their corresponding readiness attributes are depicted in Table 4.

The relationship strength test conducted also determined that the Knowledge Acquisition, Individual Interpretation, and Interface Acceptance attributes contributed to the measurement of handheld application usage readiness with relationship weight values of 0.483, 0.395, and 0.454, respectively. Each of the attributes were labelled w_{RDN1}, w_{RDN2}, and w_{RDN3} respectively, as results found that Knowledge Acquisition, Individual Interpretation, and Interface Acceptance contributed first, second and third towards the measurement and assessment of

Table 5 Weight value and symbol of readiness attributes

Goal	Attribute	Value	Symbol
Readiness	Knowledge Acquisition	0.483	w_{RDN1}
	Individual Interpretation	0.395	w_{RDN2}
	Interface Acceptance	0.454	w_{RDN3}

handheld application usage readiness *RDN*. In addition, all three readiness attributes were found to exhibit a low relationship weight value in assessing the readiness of handheld application usage. The results of the relationship test for each attribute towards assessing the readiness of handheld application usage are presented in Table 5.

2.3 Code Specificity

Formula for calculating the readiness of handheld applications usage could also be constructed by applying weights. Weight values were coded either as w_{ATTm} for representing weight value of metric, or w_{RDNa} for representing weight value of attribute. The generic symbol w_{ATTm} represents the weight code of metric mth that contributes towards its corresponding attribute ATT. Meanwhile, symbol w_{RDNa} represents the weight code of attribute ath that contributes towards measuring the overall readiness of handheld application usage, RDN.

2.4 Optimisation of Parameters

Lists of codes were produced to represent each readiness metric and attribute towards assessing the readiness of handheld application usage, presented as $M_m \bullet A_a \bullet C_{RDN}$ and $A_a \bullet C_{RDN}$, respectively. *M* represents metric, meanwhile *A* represents attribute, whereas *C* represents readiness as the goal for assessing the handheld application usage. Based on the rank order for each metric towards its corresponding attribute, presented as $M_m \bullet A_a \bullet C_{RDN, m}$ represents the sequential series (*m*th) of the metric, such as 1, 2, ..., *m*, that contributed towards a particular attribute, *a*, in assessing the handheld application usage readiness. In addition, presented as $A_a \bullet C_{RDN}$, *a* represents the sequential series (*a*th) of the attribute, such as 1, 2, ..., *a*, that contributed towards the goal for assessing the handheld application usage readiness, in which *RDN* represents the abbreviation of readiness.

Linearity code was developed based on previous work [8] regarding the identification and determination of the rank order of metrics and attributes towards assessing the handheld application usage readiness, as presented in Fig. 1. As derived from the study, results revealed that the Skills Integrated, Problems Solved, Meanings Understood and Reasons Explained metrics were ranked first ($m = 1$),

Table 6 Linearity code of readiness metrics

Attribute	Metric	Code
Knowledge Acquisition	Skills Integrated *(number of skills integrated correctly)*	$M_1 \cdot A_1 \cdot C_{RDN}$
	Problems Solved *(number of problems solved correctly)*	$M_2 \cdot A_1 \cdot C_{RDN}$
	Meanings Understood *(number of meanings understood correctly)*	$M_3 \cdot A_1 \cdot C_{RDN}$
	Reasons Explained *(number of reasons explained correctly)*	$M_4 \cdot A_1 \cdot C_{RDN}$
Individual Interpretation	Answers Supplied *(number of answers supplied correctly)*	$M_1 \cdot A_2 \cdot C_{RDN}$
	Examples Provided *(number of examples provided correctly)*	$M_2 \cdot A_2 \cdot C_{RDN}$
	Questions Given *(number of questions given correctly)*	$M_3 \ A_2 \ C_{RDN}$
Interface Acceptance	Information Memorized *(amount of information memorized correctly)*	$M_1 \cdot A_3 \cdot C_{RDN}$
	Representations Recalled *(number of representations recalled correctly)*	$M_2 \cdot A_3 \cdot C_{RDN}$
	Layouts Recognized *(number of layouts recognized correctly)*	$M_3 \cdot A_3 \cdot C_{RDN}$

second ($m = 2$), third ($m = 3$), and fourth ($m = 4$), which together contribute towards the Knowledge Acquisition attribute ($a = 1$) in assessing the readiness of handheld application usage (C_{RDN}), with linearity codes of $M_1 \cdot A_1 \cdot C_{RDN}$, $M_2 \cdot A_1 \cdot C_{RDN}$, $M_3 \cdot A_1 \cdot C_{RDN}$, and $M_4 \cdot A_1 \cdot C_{RDN}$ respectively. The Answers Supplied, Examples Provided, Questions Given metrics were coded in the order $M_1 \cdot A_2 \cdot C_{RDN}$, $M_2 \cdot A_2 \cdot C_{RDN}$, and $M_3 \cdot A_2 \cdot C_{RDN}$. It contributed towards the Individual Interpretation attribute ($a = 2$), ranked first ($m = 1$), second ($m = 2$), and third ($m = 3$) in assessing the handheld application usage readiness, labelled *RDN*.

Regarding the three metrics contributing towards the Interface Acceptance attribute ($a = 3$) in the assessment of handheld application usage readiness (C_{RDN}), Information Memorized ranked first ($m = 1$), Representations Recalled second ($m = 2$), and Layouts Recognized third ($m = 3$), and were therefore coded as $M_1 \cdot A_3 \cdot C_{RDN}$, $M_2 \cdot A_3 \cdot C_{RDN}$, and $M_3 \cdot A_3 \cdot C_{RDN}$. The linearity code of each readiness metric towards its corresponding attribute in assessing the readiness of handheld application usage is depicted in Table 6. Finally, the attributes Knowledge Acquisition coded as $A_1 \cdot C_{RDN}$, Individual Interpretation ($A_2 \cdot C_{RDN}$) and Interface Acceptance ($A_3 \cdot C_{RDN}$) ranked first ($a = 1$), second ($a = 2$), and third ($a = 3$), respectively, in the assessment of handheld application usage readiness, labelled as C_{RDN}. The linearity code of each readiness attribute is shown in Table 7.

An equation of readiness metric was formulated to determine the relative quantification of a target activity in comparison to a reference activity. The readiness metric expression ratio ($M_m \cdot A_a \cdot C_{RDN}$) of a target activity is calculated based

Table 7 Linearity code of readiness attributes

Goal	Attribute	Code
Readiness	Knowledge Acquisition	$A_1 \bullet C_{RDN}$
	Individual Interpretation	$A_2 \bullet C_{RDN}$
	Interface Acceptance	$A_3 \bullet C_{RDN}$

on the number of activities performed correctly (E_{target}), where the deviation is the difference between an actual activity and an expected activity ($\Delta_{target\ (actual\ -\ expected)}$). This was expressed in comparison to a reference activity calculated based on the total number of activities performed ($E_{reference}$) and the total number of expected activities (*Treference (expected)*).

Equation 1 shows a mathematical model of relative expression ratio in quantifying readiness metrics. The ratio is expressed as an expected versus actual target activity, in comparison to a reference expected activity. E_{target} is the observed readiness metric of target activity transcript, $E_{reference}$ is the observed readiness metric of reference activity transcript, *target* is the deviation of actual—expected of the target activity transcript, and T *reference* is the total of expected reference activity transcript. The expected activity could be a constant and a regulated transcript, which means that for the calculation of readiness metric ratio ($M_m \bullet A_a \bullet C_{RDN}$), the individual target expected activity, *target(expected)* and the reference expected activity, *reference(expected)* of the investigated transcript must be known, and only dependent on the target actual activity *target(actual)*.

$$= \left[\frac{\left(E_{target} \right)^{\Delta\ target(actual\ -\ expected)}}{\left(E_{reference} \right)^{T\ reference(expected)}} \right] \tag{1}$$

Table 8 shows detailed representation for quantifying the readiness metrics that contribute towards its corresponding attribute in measuring the readiness of handheld application usage.

An equation for assessing readiness attributes of handheld application usage was formulated by determining the relative summation of the product of weight and value in comparison to the average of weight. In detail, the attribute expression ratio ($A_a \bullet C_{RDN}$) of handheld application usage readiness is calculated based on the total product of each metric weightage (w_{ATTm}) multiplied by the corresponding metric values ($M_m \bullet A_a \bullet C_{RDN}$), and expressed in comparison to the average weightage of metric (w_{ATTm}). Equation 2 shows a mathematical model of relative expression ratio in quantifying readiness attributes of handheld application usage.

$$= \left[\frac{\sum_{m=1}^{M} w_{ATTm} \left(M_m \cdot A_a \cdot C_{RDN} \right)}{\left[\sum_{m=1}^{M} w_{ATTm} \right]} \right] \tag{2}$$

Table 8 Representation of readiness metrics

	Representation
Knowledge Acquisition	
Skills Integrated $(M_1 \cdot A_1 \cdot C_{RDN})$	$= \left[\dfrac{\Delta\, skills\ integrated\ correctly(actual - expected)\left(\mathrm{E}_{skills\ integrated\ correctly}\right)}{\mathrm{T}\, skills\ integrated(expected)\left(\mathrm{E}_{skills\ integrated}\right)} \right]$
Problems Solved $(M_2 \cdot A_1 \cdot C_{RDN})$	$= \left[\dfrac{\Delta\, problems\ solved\ correctly(actual - expected)\left(\mathrm{E}_{problems\ solved\ correctly}\right)}{\mathrm{T}\, problems\ solved(expected)\left(\mathrm{E}_{problems\ solved}\right)} \right]$
Meanings Understood $(M_3 \cdot A_1 \cdot C_{RDN})$	$= \left[\dfrac{\Delta\, meanings\ understood\ correctly(actual - expected)\left(\mathrm{E}_{meanings\ understood\ correctly}\right)}{\mathrm{T}\, meanings\ understood(expected)\left(\mathrm{E}_{meanings\ understood}\right)} \right]$
Reasons Explained $(M_4 \cdot A_1 \cdot C_{RDN})$	$= \left[\dfrac{\Delta\, reasons\ explained\ correctly(actual - expected)\left(\mathrm{E}_{reasons\ explained\ correctly}\right)}{\mathrm{T}\, reasons\ explained(expected)\left(\mathrm{E}_{reasons\ explained}\right)} \right]$
Individual Interpretation	
Answers Supplied $(M_1 \cdot A_2 \cdot C_{RDN})$	$= \left[\dfrac{\Delta\, answers\ supplied\ correctly(actual - expected)\left(\mathrm{E}_{answers\ supplied\ correctly}\right)}{\mathrm{T}\, answers\ supplied(expected)\left(\mathrm{E}_{answers\ supplied}\right.} \right]$
Examples Provided $(M_2 \cdot A_2 \cdot C_{RDN})$	$= \left[\dfrac{\Delta\, examples\ provided\ correctly(actual - expected)\left(\mathrm{E}_{examples\ provided\ correctly}\right)}{\mathrm{T}\, examples\ provided(expected)\left(\mathrm{E}_{examples\ provided}\right)} \right]$
Questions Given $(M_3 \cdot A_2 \cdot C_{RDN})$	$= \left[\dfrac{\Delta\, questions\ given\ correctly(actual - expected)\left(\mathrm{E}_{questions\ given\ correctly}\right)}{\mathrm{T}\, questions\ given(expected)\left(\mathrm{E}_{questions\ given}\right)} \right]$
Interface Acceptance	
Information Memorized $(M_1 \cdot A_3 \cdot C_{RDN})$	$= \left[\dfrac{\Delta\, information\ memorized\ correctly(actual - expected)\left(\mathrm{E}_{information\ memorized\ correctly}\right)}{\mathrm{T}\, information\ memorized(expected)\left(\mathrm{E}_{information\ memorized}\right)} \right]$
Representations Recalled $(M_2 \cdot A_3 \cdot C_{RDN})$	$= \left[\dfrac{\Delta\, representations\ recalled\ correctly(actual - expected)\left(\mathrm{E}_{representations\ recalled\ correctly}\right)}{\mathrm{T}\, representations\ recalled(expected)\left(\mathrm{E}_{representations\ recalled}\right)} \right]$
Layouts Recognized $(M_3 \cdot A_3 \cdot C_{RDN})$	$= \left[\dfrac{\Delta\, layouts\ recognized\ correctly(actual - expected)\left(\mathrm{E}_{layouts\ recognized\ correctly}\right)}{\mathrm{T}\, layouts\ recognized(expected)\left(\mathrm{E}_{layouts\ recognized}\right)} \right]$

Equation 2.1 shows detailed representation for quantifying readiness attribute Knowledge Acquisition $(A_1 \cdot C_{RDN})$ that contributes towards assessing the readiness (C_{RDN}) of handheld application usage as

$$= \left[\frac{\sum\limits_{m=1}^{4} w_{KAm}(M_m \cdot A_1 \cdot C_{RDN})}{\left[\sum\limits_{m=1}^{4} w_{KAm} \right]} \right] \qquad (2.1)$$

This involved the proportion of the accumulated product of weight and value of each contributed readiness metric (i.e., Skills Integrated $[(w_{KA1}) \times (M_1 \cdot A_1 \cdot C_{RDN})]$, Problems Solved $[(w_{KA2}) \times (M_2 \cdot A_1 \cdot C_{RDN})]$, Meanings Understood $[(w_{KA3}) \times (M_3 \cdot A_1 \cdot C_{RDN})]$, and Reasons Explained $[(w_{KA4}) \times (M_4 \cdot A_1 \cdot C_{RDN})]$) that contributed towards attribute Knowledge Acquisition $(A_1 \cdot C_{RDN})$ divided by the total metric weights $([(w_{KA1} + w_{KA2} + w_{KA3} + w_{KA4})]$.

Equation 2.2 shows detailed representation for quantifying the readiness attribute Individual Interpretation ($A_2 \bullet C_{RDN}$) that contributes towards assessing the readiness of handheld application usage (C_{RDN}) as

$$= \left[\frac{\displaystyle\sum_{m=1}^{3} w_{IIm}(M_m \cdot A_2 \cdot C_{RDN})}{\left[\displaystyle\sum_{m=1}^{3} w_{IIm}\right]} \right] \tag{2.2}$$

This involved the proportion of the accumulated product of weight and value of each contributed readiness metrics (i.e., Answers Supplied [$(w_{II1}) \times (M_1 \bullet A_2 \bullet C_{RDN})$], Examples Provided [$(w_{II2}) \times (M_2 \bullet A_2 \bullet C_{RDN})$], and Questions Given [$(w_{II3}) \times (M_3 \bullet A_2 \bullet C_{RDN})$]) that contributed towards attribute Individual Interpretation ($A_2 \bullet C_{RDN}$) divided by the total metric weights ([$(w_{II1} + w_{II2} + w_{II3})$].

Equation 2.3 shows detailed representations for quantifying readiness attribute Interface Acceptance ($A_3 \bullet C_{RDN}$) that contributes towards assessing the readiness of handheld application usage (C_{RDN}) as:

$$= \left[\frac{\displaystyle\sum_{m=1}^{3} w_{IAm}(M_m \cdot A_3 \cdot C_{RDN})}{\left[\displaystyle\sum_{m=1}^{3} w_{IAm}\right]} \right] \tag{2.3}$$

This involved the proportion of the accumulated product of weight and value of each contributed readiness metrics (i.e., Information Memorized [$(w_{IA1}) \times (M_1 \bullet A_3 \bullet C_{RDN})$], Representations Recalled [$(w_{IA2}) \times (M_2 \bullet A_3 \bullet C_{RDN})$], and Layouts Recognized [$(w_{IA3}) \times (M_3 \bullet A_3 \bullet C_{RDN})$]) that contributed towards attribute Interface Acceptance ($A_3 \bullet C_{RDN}$) divided by the total metric weights ([$(w_{IA1} + w_{IA2} + w_{IA3})$].

As a result of quantifying the metrics and attributes, a mathematical model for quantifying the overall readiness of handheld application usage was developed. An equation for assessing readiness of handheld application usage was formulated by determining the relative summation of the product of weight and value in comparison to the average of weight. In detail, the readiness expression ratio (C_{RDN}) of handheld application usage is calculated based on the total product of each attribute weightage (w_{RDNa}) multiplied by the corresponding attribute values ($A_a \bullet C_{RDN}$), and expressed in comparison to the average weightage of attribute (w_{RDNa}). Equation 3 shows a mathematical model of relative expression ratio in quantifying readiness of handheld application usage.

$$= \left[\frac{\displaystyle\sum_{a=1}^{A} w_{RDNa}(A_a \cdot C_{RDN})}{\left[\displaystyle\sum_{a=1}^{A} w_{RDNa}\right]} \right] \tag{3}$$

Equation 3.1 shows detailed representation for quantifying the readiness of handheld application usage (C_{RDN}) as

$$= \left[\frac{\sum\limits_{a=1}^{3} w_{RDNa}(A_a \cdot C_{RDN})}{\left[\sum\limits_{a=1}^{3} w_{RDNa} \right]} \right] \tag{3.1}$$

This involved the proportion of the accumulated product of weight and value of each contributed readiness attribute (i.e., Knowledge Acquisition [(w_{RDN1}) × (A_1 • C_{RDN})], Individual Interpretation [(w_{RDN2}) × (A_2 • C_{RDN})], and Interface Acceptance [(w_{RDN3}) × (A_3 • C_{RDN})]) that contributed towards assessing the readiness of handheld application usage (C_{RDN}) divided by the total value of readiness weights ([(w_{RDN1} + w_{RDN2} + w_{RDN3})].

2.5   Readiness Measurement Model (RMM)

As a result of these quantification methods, a model, namely Readiness Measurement Model (RMM), has been proposed which suggests how handheld application usage readiness should be evaluated. The model is organized by metrics, attributes and readiness as the goal. For each attribute, the model describes relevant readiness metrics appropriate for measurement and potential evaluation measures. The classification scheme in Fig. 2 summarizes the construct and the measures proposed throughout this research, and advance the readiness evaluation by providing a quantitative approach in assessing the readiness of handheld application usage.

The existence of interrelations between metrics and attributes should be taken into account in determining the level of readiness of handheld applications. Due to the linear and hierarchical structure of the RMM, any changes to metrics will result in changes to the attributes and consequently on the overall readiness of the handheld applications usage. For example, a low score on the matric (i.e. Skills Integrated (M_1 • A_1 • C_{RDN})) will directly affect the score of the attribute Knowledge Acquisition (A_1 • C_{RDN}) and results in significant implications for the overall readiness (C_{RDN}) of handheld applications usage, and vice versa.

However, to obtain the precise numeric value is as tangible as the likelihood of occurrences is impossible. Fortunately, exact figures for measuring readiness are not needed since the numbers are mostly used for comparison purpose only. Thus, prioritizing the readiness can be done by converting the values into words or sentences with which the evaluator from various background and understanding can interpret the information accurately and comprehensively.

Fig. 2 Readiness
Measurement Model (RMM)

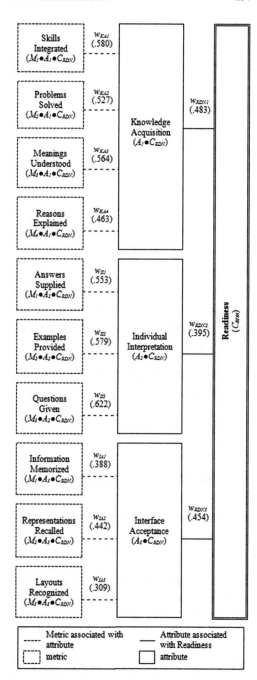

Table 9 Analysis representation of overall usage readiness

Score	Level	Status	Description
$C_{RDN} < 0.200$	1	Worst	Most badly absence or shortage of a desirable usage readiness that results users unable to perform comprehensively
$C_{RDN} < 0.400$	2	Inadequate	Lack of a desirable usage readiness that results users with the least excellent to perform task
$C_{RDN} < 0.600$	3	Acceptable	Average of a desirable usage readiness that can be tolerable to be considered as good enough
$C_{RDN} < 0.800$	4	Excellent	Complete the specific requirements of a desirable usage readiness that achieves almost in a state of being practical
$C_{RDN} \leq 1.000$	5	Outstanding	Fulfilment of all requirements of desirable usage readiness that achieves very high distinction of proficiency

Prioritizing overall readiness usage can be categorized into five distinct classifications (refer Table 9). The lowest level indicates the most badly absence or shortage of a desirable usage readiness whilst the highest level represents outstanding or fulfilment a desirable usage readiness with high distinction of proficiency.

It is important to note that prioritizing the level for measuring the readiness of handheld applications usage mentioned above is flexible and does not fixed to the stated figures. The scores for each level are open for customization and tailored to specific requirements according to the maturity of the handheld applications itself or based on the evaluator's wishes.

3 Case Study and Results

One important aspect of developing a model is to test the applicability with respect to the real world issues. As a result, a case study was used as it can verify the functionality and behaviour of the designed model in some real life context. The findings may conclude the applicability of the model to be implemented as a tool for measuring the readiness of handheld applications.

Three important criteria for the selection of a case study were used and these criteria are described as follows: *context of use*—case studies must be based on the context of actual users, tasks, equipment and environments; m*easures of readiness*—case studies were also chosen on the basis of the type of quantitative readiness measures that was about to be evaluated; and *comparison of studies*—case studies must involve comparisons between two or more experimental studies. These criteria were chosen as a basis of its potential to be working within evaluating the overall readiness of handheld applications usage.

A use-in-motion scenarios evaluation was conducted as a case study. This case study aims to validate the appropriateness of handheld applications usage that varies in representative of mobility. The evaluation of this study was performed either walking around or on a treadmill at a constant speed. Four groups of six participants were asked to perform one set of tasks; walking with high light brightness; walking with low light brightness; using treadmill with high light brightness; or using treadmill with low light brightness. Participants were assigned to perform two tasks on a stylus input. The first involved reading comprehension and the second intended to be representative of a seek-and-find task.

Two readiness measures were recorded for the reading comprehension task (i.e. number of examples provided, and number of answers supplied). In order to evaluate reading comprehension task, participants were asked to read passages via a storybook-liked application and instructed to answer a total number of twenty-five questions in 15 min treatment. A par value of expected number of answers correctly provided by participants was set to be 20 % out of the total number of questions provided. Participants were also instructed to provide example for each of the questions in 10-minute treatment and a par value of expected number of correct examples provided by participants was set to be 15 % out of the total number of questions provided.

As for evaluating the searching task, firstly participants were given a number of 20 riddles and asked to search and highlight the answers (in word) on the same storybook-liked application in 15 min. Second, participants were required to search lists of 30 objects in the pictures and highlight the hidden objects in 5 min treatment. Two measures were recorded for the searching task (number of problems solved, and number of information memorized). The expected par value for both treatments was set to be 15 %. Table 10 shows the data for each experimental condition, comparing treadmill with walking scenario and high and low brightness.

The summarization of metrics, attributes, and overall readiness scores for each experimental condition, comparing treadmill with walking scenario and high and low brightness is shown in Table 11. Results from the experiment showed that the average score of examples provided by participants using treadmill with low light brightness was slightly higher if compared to participants using treadmill with high light brightness and both low and high light brightness for walking scenario. As can be observed in Table 11, the average score of information memorized by participants using treadmill with low light brightness was also higher if compared to participants using treadmill with high light brightness and both brightness for walking scenario. However, both the ability of participants to supply answers and solve problems, the average scores were significantly higher for walking scenario in low brightness.

The analysis of the readiness attribute Individual Interpretation found that the score for performing comprehension task while walking around with low brightness was slightly higher than using treadmill with the same brightness. Results also found that walking around with high brightness was a higher than using treadmill with high brightness. The scores for each readiness attribute Knowledge Acquisition and Interface Acceptance remained the same, as Problems Solved was the only

Table 10 Collection of experimental data

Scenario	Treadmill		Walking	
Brightness	High	Low	High	Low
Comprehension Task				
T expected examples provided	25	25	25	25
Actual correct examples provided	20.83[a]	22.83[a]	21.17[a]	22.5[a]
Expected correct examples provided	3.75[b]	3.75[b]	3.75[b]	3.75[b]
Δ examples provided correctly	17.08	19.08	17.42	18.75
T expected answers supplied	25	25	25	25
Actual correct answers supplied	19.17[a]	21.83[a]	21.17[a]	22.5[a]
Expected correct answers supplied	5[b]	5[b]	5[b]	5[b]
Δ answers supplied correctly	14.17	16.83	16.17	17.5
Searching Task				
T expected problems solved	20	20	20	20
Actual correct problems solved	16.33[a]	16.83[a]	16.5[a]	17.17[a]
Expected correct problems solved	3[b]	3[b]	3[b]	3[b]
Δ problems solved correctly	13.33	13.83	13.5	14.17
T expected information memorized	30	30	30	30
Actual correct information memorized	20.33[a]	24.17[a]	20.33[a]	24.0[a]
Expected correct information memorized	4.5[b]	4.5[b]	4.5[b]	4.5[b]
Δ information memorized correctly	15.83	19.67	15.83	19.5

[a]Mean scores of target activity, [b]Par value of target activity

metric that corresponded to attribute Knowledge Acquisition whilst Information Memorized was the only metric that corresponded to attribute Interface Acceptance. Therefore, results concluded that participants can concentrate more on the software application to obtain more knowledge while walking with low light brightness. Meanwhile, participants accepted more on the interface of the software application while on the treadmill with low light brightness.

The final scores indicate that the handheld application used in the study completes the specific requirements of a desirable usage readiness that achieves almost in a state of being practical for the comprehension task. However, comparing to the experimental scenarios thus confirmed that the readiness of handheld application usage is significantly higher by walking around than using treadmill at a constant speed. Results also found that there are significant effects between different brightness where the readiness of handheld application usage is significantly higher with low brightness compared to the high light brightness.

The searching task scores also indicate that the handheld application used in the study completes the specific requirements of a desirable usage readiness that achieves almost in a state of being practical for the walking scenarios. However, comparing to the lighting brightness thus confirmed that the readiness of handheld application usage is significantly lower by using treadmill in high light brightness which can be considered good enough to be used in high brightness for searching

Table 11 Metrics, attributes and overall readiness scores

		Comprehension Task		Searching Task	
Scenario	Brightness	Examples Provided $(M_2 \cdot A_2 \cdot C_{RDN})$	Answers Supplied $(M_1 \cdot A_2 \cdot C_{RDN})$	Problems Solved $(M_2 \cdot A_1 \cdot C_{RDN})$	Information Memorized $(M_1 \cdot A_3 \cdot C_{RDN})$
Treadmill	High	0.6832	0.5668	0.6665	0.5277
	Low	0.7632	0.6732	0.6915	0.6557
Walking	High	0.6968	0.6468	0.6750	0.5943
	Low	0.7500	0.7000	0.7085	0.6500
		Comprehension		Searching	
Scenario	Brightness	Individual Interpretation $(A_2 \cdot C_{RDN})$		Knowledge Acquisition $(A_1 \cdot C_{RDN})$	Interface Acceptance $(A_3 \cdot C_{RDN})$
Treadmill	High	0.6263[a]		0.6665[b]	0.5277[c]
	Low	0.7193[a]		0.6915[b]	0.6557[c]
Walking	High	0.6723[a]		0.6750[b]	0.5943[c]
	Low	0.7256[a]		0.7085[b]	0.6500[c]
		Comprehension		Searching	
Scenario	Brightness	Readiness (C_{RDN})		Readiness (C_{RDN})	
Treadmill	High	0.6263[d]		0.5993[e]	
	Low	0.7193[d]		0.6742[e]	
Walking	High	0.6723[d]		0.6359[e]	
	Low	0.7256[d]		0.6801[e]	

[a]$[((w_{II2}) \times (M_2 \cdot A_2 \cdot C_{RDN}) + (w_{III}) \times (M_1 \cdot A_2 \cdot C_{RDN}))/(w_{II2} + w_{III})] = [((0.579) \times (M_2 \cdot A_2 \cdot C_{RDN}) + (0.553) \times (M_1 \cdot A_2 \cdot C_{RDN})/(0.579 + 0.553)]$, [b]$[(w_{KA2}) \times (M_2 \cdot A_1 \cdot C_{RDN})/(w_{KA2})] = [(0.527) \times (M_2 \cdot A_1 \cdot C_{RDN})/(0.527)]$, [c]$[(w_{IA1}) \times (M_1 \cdot A_3 \cdot C_{RDN})/(w_{IA1})] = [(0.388) \times (M_2 \cdot A_1 \cdot C_{RDN})/(0.388)]$, [d]$[(w_{RDN2}) \times (A_2 \cdot C_{RDN})/(w_{RDN2})] = [(0.395) \times (M_2 A_1 \cdot C_{RDN})/(0.395)]$, [e]$[((w_{RDN1}) \times (A_1 \cdot C_{RDN}) + w_{RDN3}) \times (A_3 \cdot C_{RDN}))/(w_{RDN1} + w_{RDN3})] = [((0.483) \times (A_1 \cdot C_{RDN}) + (0.454) \times (A_3 C_{RDN})/(0.483 + 0.454)]$

Table 12 Analysis of overall readiness

Scenario	Brightness	Comprehension	Searching
Treadmill	High	Excellent	Acceptable
	Low	Excellent	Excellent
Walking	High	Excellent	Excellent
	Low	Excellent	Excellent

task. Table 12 shows the summarization of the analysis of each experimental condition by comparing treadmill versus walking scenario and high versus low brightness.

4 Conclusion

Both the theory and practice of quantifying handheld application usage readiness have been hampered by the absence of a thorough mathematically based model as a method for evaluation. As a result, this research effort has been in a position to derive a preliminary mathematically based specification and measurement scheme specifically for assessing the handheld application usage readiness. The ultimate value for developing a mathematical oriented approach is to provide a systematic and quantitative method for conducting handheld application usage readiness evaluation research.

References

1. Thakur, R., Srivastava, M.: Adoption readiness, personal innovativeness, perceived risk and usage intention across customer groups for mobile payment services in India. Internet Res. **24** (3) (2014)
2. Cheon, J., Lee, S., Crooks, S.M., Song, J.: An investigation of mobile learning readiness in higher education based on the theory of planned behaviour. Comput. Educ. **59**(3), 1054–1064 (2012)
3. Abas, Z.W., Peng, C.L., Mansor, N.: A study on learner readiness for mobile learning at Open University Malaysia. In: Proceedings of the International Conference Mobile Learning (IADIS 2009), pp. 151–157 (2009)
4. Andaleeb, A.A., Idrus, R.M., Ismail, I., Mokaram, A.K.: Technology readiness index (TRI) among USM distance education students according to age. Int. J. Hum. Soc. Sci. **5**(3), 189–192 (2010)
5. Hussin, S., Manap, M.R., Amir, Z., Krish, P.: Mobile learning readiness among Malaysian students at Higher Learning Institutes. In: Proceedings of the APAC MLEARNING Conference (2011)
6. Lu X, Viehland D. Factors influencing the adoption of mobile learning. In: Proceedings of the 19th Australasian Conference on Information Systems, pp. 597–606 (2008)
7. Cheung, S.K.S., Yuen, K.S., Tsang, E.Y.M.: A study on the readiness of mobile learning in open education. In: Proceedings of the International Symposium on IT in Medicine and Education (ITME'2011), pp. 133–136 (2011)
8. Fadzlah, A.F.A.: Identifying measures for assessing the readiness of handheld application usage. Lect. Notes Softw. Eng. **2**(3), 256–261 (2014)
9. Ziarati, K., Khayami, R., Parvinnia, E., Milani, G.A.: Virtual collaboration readiness measurement a case study in the automobile industry. In: Adv. Comput. Sci. Eng. pp. 913–916 (2009)
10. Said, R.F.M., Rahman, S.A., Mutalib, S., Yusoff, M., Mohamed, A.: User technology readiness measurement in fingerprint adoption at higher education institution. In: Comput. Sci. Its Appl.–ICCSA, 91–104 (2008)
11. Paver, N.C., Khan, M.H., Aldrich, B.C., Emmons, C.D.: Accelerating mobile video: a 64-Bit SIMD architecture for handheld applications. J. VLSI Sig. Process. Syst. Sig. Image Video Technol. **41**(1), 21–34 (2005)

12. Caban-Martinez, A.J., Clarke, T.C., Davila, E.P., Fleming, L.E., Lee, D.J.: Application of handheld devices to field research among underserved construction worker populations: a workplace health assessment pilot study. Environ. Health **10**(1), 1–5 (2011)
13. Roth, J., Unger, C.: Using handheld devices in synchronous collaborative scenarios. Pers. Ubiquit. Comput. **5**(4), 243–252
14. Thomas, P., Gellersen, H.: Handheld and Ubiquitous Computing. Springer Berlin
15. Chunlin, L., Layuan, L.: A market-based mechanism for integration of mobile devices into mobile grids. Int. J. Ad Hoc Ubiquit. Comput. 01–07 (2010)

Triple Extraction Using Lexical Pattern-based Syntax Model

Ai Loan Huynh, Hong Son Nguyen and Trong Hai Duong

Abstract This work proposed a new approach to extract relations and their arguments from natural language text without knowledge base. Using the grammar of English language, it allows detecting sentence based on verb types and phrasal verb in terms of extraction. In addition, this approach is able to extract the properties of objects/entities mentioned in text corpus, which previous works have not yet explored. Experimental result is performed by using various real-world datasets which were used by ClausIE and Ollie, and other text were found in the Internet. The result shows that our method is significant in comparison with ClausIE and Ollie.

Keywords Tripe extraction · Semantics · Open information extraction · Relations extraction

1 Introduction

Today, the fast growth of Web as well as the variety of data formats and language enables challenges for both Information Retrieval and Natural Language Processing in finding relevant documents which satisfy the needs of users. This fact leads us to the necessity of discover structured information from unstructured or semi-structured sources to retrieve information. Indeed, the growth demand of searching systems

A.L. Huynh · T.H. Duong (✉)
School of Computer Science and Engineering, International University,
VNU-HCMC, Ho Chi Minh City, Vietnam
e-mail: haiduongtrong@gmail.com

A.L. Huynh
e-mail: ailoan2712@gmail.com

H.S. Nguyen
Faculty of Information Technology, Le Quy Don University, Hanoi, Vietnam
e-mail: son_nguyenhong2002@yahoo.com

© Springer International Publishing Switzerland 2016
T.B. Nguyen et al. (eds.), *Advanced Computational Methods
for Knowledge Engineering*, Advances in Intelligent Systems
and Computing 453, DOI 10.1007/978-3-319-38884-7_19

induces the development of Information Extraction (IE) systems that analyze text written in plain natural language and to find facts or events in the text.

In IE system, its methods can be used to build knowledge representation models that report relations between words like ontology, semantic network, etc. The traditional IE systems learnt an extractor for each target relation as an input from labeled training examples [7, 10, 11]. Moreover, this approach did not scale to corpora where the number of target relations is very large, or where the target relations cannot be specified in advance [5].

Aiming to overcome the above problem, the Open Information Extraction (Open IE) where relation phrases are identified was introduced with Text Runner system [1]. Open IE systems extracted a large number of triples (*arg1, rel, arg2*) from text based on verb-based relations, where *arg1* and *arg2* are the arguments of the relation and *rel* is a relation phrase. Unlike other relation extraction methods which focused on a pre-defined a set of target relations, Open IE systems based on unsupervised extraction methods. They did not require any background knowledge or manually labeled training data.

The two tools that made up the Open IE state-of-the art, Ollie [9] and ClausIE [3], which intended to export the largest number of relations from the same sentence. They generated triples which are near to reproductions of the text. However, they lost the minimality of triple as mentioned in CSD system [2]. Because *Subject* or *Object* may consist of one or more entities in the same relation, Ollie and ClausIE did not identify the fact of each entity with its relation. For instance, a give sentence "*Anna and Jack have a meeting*". Two entities "*Anna*" and "*Jack*" refer to the relation "*have a meeting.*" A correct decomposition of this sentence would yield the triples "*Anna, have, a meeting*" and "*Jack, have, a meeting*"

The recent system LSOE [14] improves the precision of previous Open IE. It performs rule-based extraction of triples using POS-tagged text by applying lexical syntactic patterns based on Pustejovsky's qualia structure and generic patterns for non-specified relationships. But it takes the limitation of qualia-based patterns. In addition, LSOE as well as CSD use POS-tagged as the input.

Another issue is due to Verb phrase (VP) parsing. Most of Open IE systems relied on verb-based relation. The starting point of a VP in a clause is recognized but the VP is not fully parsed. VP words are simply connected together to create relations so that OIE methods create several relations that have the same meaning but different grammar forms, such as "is the author of" and "are the authors of"; "is", "are", "was", "were" and "has been"; "have", "has", "had"... Voice and affirmative of verbs are also not processed. Phrasal verbs often require complex grammar structures.

The motivation for our approach comes from the grammar structure of a sentence in English as well as verb-phrase patterns. In this paper, according to the approach presented in [8], a triple in a sentence expressed the relation between subject and object, in which the relation is as a verb phrase. The goal of the proposed algorithm is to extract the sets of triple with form {*Subject, Verb-Phrase, Object*} out of syntactically parsed sentences based on Syntax Model (SM) of

English language by determining verb usage pattern (VUP). With SM, not only the relationship between subject and object, but also the relationship between entity and its properties are pointed out in triples.

2 Related Works

In recent years, the process of Open Information Extraction has been improved in some systems such as TextRunner [1], Reverb [5], Ollie [9], ClausIE [3], CSD-IE [2] and LSOE [14]. In the following, we shortly summarize how the old systems (including Reverb, Ollie, ClausIE, CSD-IE and, LSOE) were implemented.

Reverb was a shallow extractor that reduced the incoherent extractions and uninformative extractions in previous open extractors by proposing two concepts on relation phrases including syntactic and lexical constraints. Firstly, Reverb extracted relation phrases that satisfied these two constraints, and then identified arguments as noun phrases that were positioned left and right of the extracted relations. The relation was the longest sequence of words staring from a verb satisfied all constraints.

Ollie improved the Open IE systems by addressing two important limitations of extraction quality: relations not mediated by verbs and relations expressed as a belief, attribution or other conditional context. It used Reverb to make a set of seed tuples and then bootstrapped the training set to learn "open pattern templates" that determined both the arguments and the extract relation phrase as Reverb. Ollie reached from 1.9 to 2.7 times bigger area under precision-yield curve, compared with Reverb and WOE systems.

ClausIE simply detected clauses and clause types via the dependency output from a probabilistic parser. A clause was understood as a part of a sentence, constituted by SVO (subject, verb, object) and some adverbs. Relied on dependency parser, ClausIE recognized one of 12 patterns of 5 verb types (copular verb, intransitive verb, di-transitive verb, mono-transitive verb and complex-transitive verb) to discover the clause types. The verb phrase was not parsed and used as the relation name. The results of this system were done in three datasets and presented as "the number of correct extractions/the total number of extractions": 1706/2975 for 500 sentences from Reverb dataset; 598/1001 for 200 sentences from Wikipedia; and 696/1303 for 200 sentences from New York Times.

CSD-IE decomposed a sentence into sentence constituents by defining a set of rules on the parsed tree structure manually. Then, the identified sentence constituents were combined to form the context, and extract triples from the resulted context. The authors compared their method with Reverb, ClausIE and Ollie with *New York Times* and *Wikipedia* datasets used in [3] and obtained an average of 70 % precision.

Xavier and Lima [15] extracted noun compounds and adjective-noun pairs from noun phrase, interpret extracted information by lexical-syntactic analysis and exports relations. This method enhanced the extraction of relations within the noun compounds and adjective-noun pairs which is the gap in Open IE.

LSOE was the most recent OIE system which used POS-tagged text as input and applies a pattern-matching technique with using lexical syntactic patterns. It defined two kinds of patterns: generic patterns to identify domain specific non-specified relations and Qualia-based patterns. The weaknesses of LSOE were about the limit number of qualia-based patterns and generic patterns defined manually. Therefore, it missed the extraction of relations expressed by verbs. The authors reported that LSOE extracted less tuples than Reverb, but it achieved 54 % of precision while Reverb obtained 49 %.

3 Triple Extraction Using Lexical Pattern-based Syntax Model

3.1 Observation

The methodology of this approach is proposed via the analysis of English grammar structure. As same as Open IE systems like ClausIE or Ollie, which uses verb-based relation to extract triples, the new approach also depends on the verb type or the phrasal verb in use in order to transfer a sentence into its syntax model.

Considering some English sentences as below:

From three above tables Tables 1, 2 and 3, it is given that a sentence has at least one main clause. Each clause is constituted by subject, verb and objects. A subject can be either noun phrases or clause. An object can be a noun phrase, an adjective phrase, or a clause in Table 4.

A noun phrase starts with a head noun. It may contain pre-modifier or post-modifier of this head noun. Pre-modifier can be a noun, an adjective phrase or a participle. Post-modifier can be a subordinate clause or prepositional phrase (shown in Table 5).

Table 1 The grammar structure of a simple sentence

A simple sentence		
He	Is	Handsome
Subject	**Verb**	**Object**

Table 2 The grammar structure of a sentence with multiple clauses

A sentence with multiple clauses						
I	Have	a shirt	;	It	Is	Beautiful
Subject	Verb	Object		Subject	Verb	Object
Clause 1				**Clause 2**		

Table 3 The grammar structure of a complex sentence

A complex sentence is constituted by subordinate clause and main clause					
Since	*we*	*can't go*	*You*	*can have*	*the tickets.*
Subordinator	Subject	Verb			
Subordinate Clause			**Main Clause**		
Adverbial			**Subject**	**Verb**	**Object**

Table 4 The grammar structure of an object as a clause

A subordinate clause functioning as object		
I	*Know*	*that you lied.*
		Subordinate Clause
Subject	**Verb**	**Direct Object**

Table 5 The grammar structure of a noun phrase

Some Examples of the Noun Phrase				
Function	*Determiner*	*Pre-modifier*	*Head*	*Post-modifier*
1			Lions	
2	The	Information	Age	
3	Several	new mystery	Books	which we recently enjoyed
4	A	Marvelous	data bank	filled with information
FORMS	*Pronoun*	*Participle*	*Noun*	*Prepositional Phrase*
	Article	*Noun*	*Noun*	*Relative Clause*
	Quantifier	*Adjective Phrase*	*Pronoun*	*Nonfinite Clause*
				Complementation

Table 6 The grammar structure of prepositional phrase

Some Examples of the Prepositional Phrase		
Function	*Preposition*	*Complement*
1	With	Her
2	On	The table
3	From	what I can see
FORMS	**Preposition**	*Adverb*
		Noun Phrase
		-ing Clause or Relative Clause

A prepositional phrase is subdivided into a preposition and a complement. In which, complement may be an adverb, a noun phrase or a subordinate clause (shown in Table 6).

Table 7 The grammar structure of adjective phrase

Some Examples of the Adjective Phrase			
Function	Pre-modifier	Head	Post-modifier
1		Happy	
2		Young	in spirit
3	Too	Good	to be true
4		Excited	Indeed
FORMS	*Adverb*	*Adjective*	***Prepositional Phrase***
			Infinitive Clause
			Adverb

An adjective phrase begins with an adjective as a head word. It also has pre-modifier and post-modifier to help the adjective word to be more clearly as given examples in Table 7.

In general, the analysis method on the grammar structure of a sentence is based on the observations in Fig. 1. These observations are concluded by examining the sentence writing of human being as in previous examples.

The simple sentence only has one clause which contains subject and predicate (*for exp: I have a book*). But the complex sentence includes many clauses which are linked together via subordinators, coordinating conjunctions or semi-colon. And the format of sentence depends on the human writing style. For instance, some writers prefer to use coordinating conjunction "and" replaced to semi-colon in order to connect multiple independent clauses of a sentence.

There are several verb types, e.g. transitive, intransitive, linking, etc. For each verb type, a grammar structure is required in usage. For example, an intransitive verb does not have a direct object but a transitive verb requires a noun or noun phrase as a direct object. However, a di-transitive verb requires two nouns or noun phrases as direct and indirect objects. A verb can have several verb types, for

---

a) A sentence can be one of the below forms:
   a1. Sentence = Subject + Predicate
   a2. Sentence = Independent Clause + coordinating conjunction + Independent Clause
   a3. Sentence = Adverbial Clause + Main Clause
b) Clause (C) = Subject (Subj) + Predicate (Pre)
c) Subject can be a noun phrase (NP) or a clause
d) Predicate = Verb + Direct Object (Dobj) + Indirect Object (Iobj) or Complement (Comp)
e) DO can be the list of NP or Clause or Pronoun
f) IO can be the list of NP or Clause or Pronoun
g) Comp can be the list of NP, list of Adjective Phrase (AdjP), an Adverb Phrase (AdvP) , a Prepositional Phrase (PrepP) or a Clause

Fig. 1 The observations

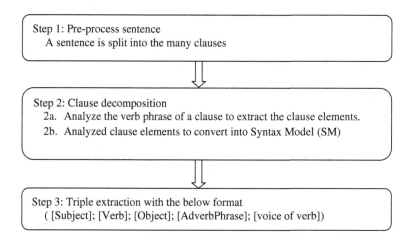

Fig. 2 Steps process a sentence

examples, "remember" can be transitive as well as intransitive verb. Therefore, as in observation (d), the predicate should follow the predefined grammar structure of that verb and its verb type.

3.2 New Triple Extraction Algorithm

Following the observation in Fig. 1, it is given that a sentence has at least a clause. In addition, depending on the verb type or phrasal verb, a verb requires a particular grammar structure called Verb Usage Pattern (VUP). Whether a verb is used in a sentence, the grammar structure of the sentence has to be satisfied the current used VUP of this verb. For instance, word *"remember"* is both transitive and intransitive verb. A sentence made by this verb has one of two VUPs: *Subject + Verb* or *Subject +Verb + Object*. For another example, the separable phrasal verb *"switch on"* has two patterns in grammar structure of the sentence: *John switches on the radio (Subject + Verb + particle + Object)* or *John switches the radio on (Subject + Verb + Object + particle).* The construction of VUP(s) is based on Internet resources from Oxford Online Dictionary,[1] the Free Dictionary[2] or collected from linguistic resources in [6, 12, 13].

In this paper, triple extraction system performs in three steps as Fig. 2. Firstly, the sentence is divided into many clauses. In the second step, each clause is processed to decompose into clause elements and convert into Syntax Model (SM) by

---

[1]www.oxforddictionaries.com.

[2]www.thefreedictionary.com.

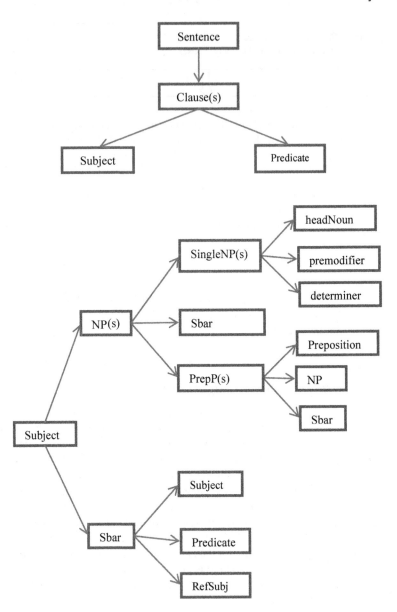

Fig. 3 Syntax model (part 1)

applying VUP. Finally, the tree constituents by clause elements are combined to extract triples.

Step 1: Sentence does pre-processing to divide into multiple clauses. As mentioned in part 3.1, the format of sentence relies on the human writing. Therefore, the sentence firstly detects whether it has multiple independent clauses via semi-colon.

Next, each independent clause plays the role as a simple sentence; and this sentence continues to check whether it has format type a3 in observation Fig. 1 through subordinators. Finally, the list of clauses is gotten to do the analysis in next steps.

Step 2: Clause Decomposition consists of two main sub-tasks: (2a) parsing a verb phrase (VP) and (2b) converting into SM

- **Sub-task 2a**: a probabilistic parser is called to produce the parsed structure, which can be a phrase structure trees or a typed dependencies. The method can work on both forms of parsed structure, but in this paper, only the Stanford typed dependencies are used to illustrate. This task starts with searching main

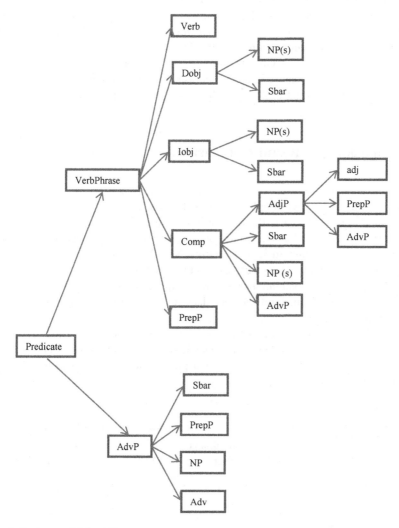

Fig. 4 Syntax model (part 2)

verb of clause based on ROOT and COP typed dependencies using Stanford Parser. Applying the defined VUP before, parts of clause such as Subject, Verb and Object are identified.

- **Sub-task 2b**: Each of extracted elements in previous task is sub-divided and mapped into smaller elements like Noun Phrase, Adjective Phrase, Prepositional Phrase … etc. as same as syntax model (SM) in Figs. 2 and 3.

Step 3: Triples with above format are extracted for each clause as well as for whole sentence (Fig. 4)

With SM, more detail triples which describe the relationship between sub-elements with each part of clause are given. For instance, a noun phrase consisting of prepositional phrase as a modifier like "*a girl with blond hair*" can also have a triple which shows the meaning in relation between these two elements. In this case, triple *{a girl, is (complemented by), with blond hair}* is given. Moreover, SM can identify more specific elements in objects compared with Ollie and ClausIE. It separates Objects into lists of noun phrase, list of preposition phrases supporting for clause or adverb phrase. Therefore, it makes enhancement in the accuracy of extracted triples as well as their minimality [2].

4 Experimental Result Comparison

We compare the proposed system against the two Open IE systems Ollie and ClausIE. This system was implemented as describe in Sect. 3, using the Stanford Constituent Parser [4]. ClausIE was run in default mode to extract triples.

Datasets used for this experience are: 200 random sentences from English Wikipedia and 200 random sentences from New York Time which are the exact same datasets as in [3]. Our results are summarized in Table 8.

Table 8 shows the total number of correct extractions as well as the total number of extractions for each method and each dataset. These result numbers are evaluated by human experience.

This algorithm achieves the better results in giving the detailed triples compared with previous Open IE systems. However it still remains some limitations in the improvement of quality and quantity of triples because of parser failure and the complexity of sentence.

Table 8 Number of correct extractions and total number of extractions

	Ollie	ClausIE	New system
New York dataset	270/500	662/926	1119/1542
Wiki dataset	357/573	602/794	1008/1323

Table 9 Result of example 1

System	Output
ClausIE	("the girl with blond hair", "is", "beautiful")
Ollie	(the girl; is; beautiful)
	(beautiful; be the girl with; blond hair)
New algorithm	{girl with blond hair; be; beautiful;; active}
	{girl; be (complemented by); with blond hair;; active}
	{blond; is property of; hair;;}

4.1 Good Points

There are two main points that new algorithm give a better result compared with Ollie and ClausIE. They will be described in some examples as below:

Example 1 "the girl with blond hair is beautiful"
Firstly, both Ollie and ClausIE did not give any triples which show the relationship between the modifier and its noun phrase, or between noun phrase and its properties. In example 1, *"blond"* is a property of *"hair"* (Table 9).

The extracted triple in this case has formats:

([adjective], "be property of", [head Noun])
([single NP], "be complemented by", [prepPhrase])

Considering two another examples:

Example 2 "LaBrocca scored his first goal for the club in a 4-1 home victory vs. Chicago Fire."

Example 3 "Jack and Janny went to school"
Secondly, the problem of ClausIE and Ollie is still missing the details of an extraction. It means that the extracted fact by both systems may be described in more facts. Our system has the same idea in improve quality of extraction in the minimality of the extracted facts. It is proved in example 2 and example 3 obviously (Tables 10 and 11).

4.2 Weak Points

Mausam et al. [9] and Del Corro and Gemulla [3] claimed that most of the extraction errors are due to two problems: parser failures and inability to express relationships in the text into binary relations. And this approach also takes the mistakes related to these issues:

Table 10 Result of example 2

System	Output
ClauseIE	("LaBrocca", "scored", "his first goal for the club")
	("LaBrocca", "scored", "his first goal in a 4-1 home victory vs. Chicago Fire")
	("LaBrocca", "scored", "his first goal")
	("his", "has", "first goal")
Ollie	(LaBrocca; scored for; the club)
	(LaBrocca; scored; his first goal)
New algorithm	{LaBrocca; score; his first goal;; active}
	{LaBrocca; score; for club;; active}
	{LaBrocca; score; his first goal for club;; active}
	{LaBrocca; score; in 4-1 home victory vs. Chicago Fire;; active}
	{LaBrocca; score; his first goal in 4-1 home victory vs. Chicago Fire;; active}
	{4-1 home victory; be (complemented by); vs. Chicago Fire;; active}
	{first; is property of; goal;;}
	{he; have; first goal;;}

Table 11 Result of example 3

System	Output
ClausIE	("Jack and Janny", "went", "to school")
Ollie	(Jack and Janny ; went to ; school)
New algorithm	{Jack ; go to ; school ; ; active}
	{Janny ; go to ; school ; ; active}
	{Jack, Janny ; go to ; school ; ; active}

Example 4 "Judy laughed and Jimmy cried"

Stanford Parser shows that

```
(ROOT
 (S
  (NP (NNP Judy) (NNP laughed)
   (CC and)
   (NNP Jimmy))
  (VP (VBD cried))))
nn(laughed-1, Judy-0)
nsubj(cried-4, laughed-1)
conj_and(laughed-1, Jimmy-3)
root(ROOT-0, cried-4)
```

In this example, the failure in extraction is due to the error of Stanford dependency in Stanford parser. The word *"laughed"* should be a VP instead of NNP. Therefore, the output of this sentence as below in new algorithm is not correct:

{Judy laughed ; cry ; ; ; active}
{Jimmy ; cry ; ; ; active}
{Judy laughed , Jimmy ; cry ; ; ; active}

Another issue is due to the sentence writing of human beings. This algorithm cannot cover syntax model of complex sentence which is constituted by too many clauses as the below example:

Example 5 "*Daughter of the actor Ismael Sanchez Abellan and actress and writer Ana Maria Bueno (better known as Ana Rosetti), Gabriel was born in San Fernando, Cadiz, but spent her childhood in Madrid*" (Table 12).

Table 12 Result of example 5

System	Output
ClausIE	("Gabriel", "was born", "in San Fernando")
	("Daughter of the actor Ismael Sanchez Abellan and actress and writer Ana Maria Bueno better known as Ana Rosetti", "Cadiz")
	("Daughter of the actor Ismael Sanchez Abellan and actress and writer Ana Maria Bueno better known as Ana Rosetti", "spent", "her childhood in Madrid")
	("Daughter of the actor Ismael Sanchez Abellan and actress and writer Ana Maria Bueno better known as Ana Rosetti", "spent", "her childhood")
	("her", "has", "childhood")
Ollie	(Gabriel; was born in; San Fernando)
	(her childhood; be spent in; Madrid)
	(Daughter of the actor Ismael Sanchez Abellan and actress and writer Ana Maria; was born in; San Fernando)
	(Gabriel; be known as; Ana Rosetti))
	(Gabriel; was born at; San Fernando)
	(Gabriel; was born on; San Fernando)
	(Daughter of the actor Ismael Sanchez Abellan and actress and writer Ana Maria; be known as; Ana Rosetti))
	(San Fernando; was born in; Cadiz)
	(Daughter of the actor Ismael Sanchez Abellan and actress and writer Ana Maria; was born at; San Fernando)
	(Daughter of the actor Ismael Sanchez Abellan and actress and writer Ana Maria; was born on; San Fernando)
New Algorithm	{Daughter of actor Ismael Sanchez Abellan, actress Ana Maria Bueno ; bear ; in San Fernando ; better known as Ana Rosetti ; passive}
	{San Fernando ; be ; Cadiz ; ; active}
	{actor Ismael Sanchez Abellan,actress Ana Maria Bueno ; be ; Gabriel ; ; active}
	{Daughter ; be (complemented by) ; of actor Ismael Sanchez Abellan,actress Ana Maria Bueno ; ; active}
	{Ismael Sanchez Abellan ; be ; actor ; ; active}
	{Ana Maria Bueno ; be ; actress ; ; active}

In this example, all three systems miss information as below:

(Ismael Sanchez Abellan is actor.
Ana Maria Bueno is actress.
Ana Maria Bueno is a writer.
Ana Maria Bueno is Ana Rosetti.
Gabriel was born in San Fernando, Cadiz.
Gabriel spent her childhood in Madrid.
Daughter of the actor Ismael Sanchez Abellan and actress and writer Ana
Maria Bueno is Gabriel.)

Besides, the main idea to convert a sentence into syntax model bases on the defined VUP. We cannot list full types of a verb, and then it cannot give any result in some cases.

5 Conclusion

This paper presented triple extraction based on syntax model of English language. The algorithm is implemented by determining verb usage pattern (VUP) in order to convert sentence into syntax model. The experimental result indicates that new system improved the quality of extraction in the minimality. Besides, its triples can describe the relationship between properties or modifier of a noun phrase. In this experience, that more detailed triples are extracted will enhance the number of relevant triples in sentence.

However, because of the dependence on a probabilistic parser, it leads to some issues which are due to the parser. Additionally, syntax model cannot adapt with all kinds of sentence written by human. Furthermore, the problem concerned about the performance in precision of triples should be considered. All of these issues allow us to do more researches in order to achieve the better result without knowledge-based as well as with using the defined knowledge in future.

References

1. Banko, M., Cafarella, M.J., Soderland, S., Broadhead, M., Etzioni, O.: Open Information extraction from the web. In: IJCAI, pp. 2670–2676 (2007)
2. Bast, H., Haussmann, E.: Open information extraction via contextual sentence decomposition. In: IEEE Seventh International Conference on Semantic Computing (ICSC), pp. 154–159 (2013)
3. Corro, L.D., Gemulla, R.: ClausIE: clause-based open information extraction. In: WWW, pp. 355–366 (2013)
4. de Marnee, M.-C., Manning, C.D.: Stanford typed dependencies manual
5. Etzioni, O., Fader, A., Christensen, J., Soderland, S., Mausam, M.: Open information extraction: the second generation. In: IJCAI, vol. 11, pp. 3–10 (2011)

6. Hampe, B.: Transitive phrasal verbs in acquisition and use: a view from construction grammar. Lang. Value **4**(1), 1–32 (2012)
7. Kim, J., Moldovan, D.: Acquisition of semantic patterns for information extraction from corpora. In: Proceedings. of Ninth IEEE Conference on Artificial Intelligence for Applications, pp. 171–176 (1993)
8. Leskovec, J., Grobelnik, M., Milic-Frayling, N.: Learning sub-structures of document semantic graphs for document summarization. In: Proceedings of the 7th International Multi-Conference Information Society IS, vol. B, pp. 18–25 (2004)
9. Mausam, Schmitz, M., Soderland, S., Bart, R., Etizioni, O.: Open language learning for information extraction. In: EMNLP-CoNLL, pp. 523–534 (2012)
10. Riloff, E.: Automatically constructing extraction patterns from untagged text. In: Proceedings. of the Thirteenth National Conference on Artificial Intelligence (AAAI-96), pp. 1044–1049 (1996)
11. Soderland, S.: Learning Information Extraction Rules for Semi-Structured and Free Text. Mach. Learn. **34**(1–3), 233–272 (1999)
12. Thim, S.: Phrasal verbs: The English Verb-particle Construction and its History, vol. 78. Walter de Gruyter (2012)
13. Trebits, A.: The most frequent phrasal verbs in English language EU documents–A corpus-based analysis and its implications. System **37**(3), 470–481 (2009)
14. Xavier, C.C., De Lima, V.L.S., Souza, M.: Open information extraction based on lexical semantics. J. Braz. Comput. Soc. **21**(1), 1–14 (2015)
15. Xavier, C., Lima, V.S.: Boosting open information extraction with noun-based relations. In: Proceedings of the ninth international conference on Language Resources and Evaluation (LREC'14). European Language Resources Association (ELRA), pp. 96–100. Reykjavik, Iceland (2014)

Using Mathematical Tools to Reduce the Combinatorial Explosion During the Automatic Segmentation of the Symbolic Musical Text

Michele Della Ventura

Abstract This document is going to focus on the modalities of management of the combinatorial explosion problem deriving from computer-aided music analysis: a major problem, most of all, for those who perform automatic analysis of the musical text considered at a symbolic level, due to the high number of recognizable "musical objects". While briefly introducing the results of the application of different processing techniques, this article shall discuss the necessity to define a series of procedures meant to reduce the number of final "musical objects" to use in order to identify a melody (or a musical theme), by selecting the ones that do not carry redundant information. Consequently, the results of their application shall be presented in statistic tables, in order to provide information on how to reduce the musical objects to a small number, so as to ensure major precision in the musical analysis.

Keywords Combinatorial explosion · Musical analysis · Musical object · Segmentation · Similarity

1 Introduction

One of the main objectives in the field of artificial intelligence (AI) is to develop systems able to reproduce intelligence and human behavior: the machine is not expected to be able to have the same cognitive abilities as humans [1], or to be aware of what it is doing, but only to know how to efficiently and optimally solve problems, being them difficult ones, in specific fields of action [2]. Therefore, the purpose of the studies carried out in the field of AI is not to replace human beings in all their capacities, but to support and improve human intelligence in certain specific fields [1, 3]: the improvement may be based on the computing power

M. Della Ventura (✉)
Department of Technology, Music Academy "Studio Musica", Treviso, Italy
e-mail: dellaventura.michele@tin.it

© Springer International Publishing Switzerland 2016
T.B. Nguyen et al. (eds.), *Advanced Computational Methods for Knowledge Engineering*, Advances in Intelligent Systems and Computing 453, DOI 10.1007/978-3-319-38884-7_20

281

Fig. 1 Symbolic musical text

derived from the use of computers. Many works in the artificial intelligence field highlight the importance of computational environments for supporting researchers in the organization of experimental data and in the generation, evaluation and revision of theoretical knowledge.

This article concentrates on the analysis of the musical text considered on a symbolic level (Fig. 1) and therefore, on the segmentation process that can be performed by a computer: a process the purpose of which is to identify "musical objects" [4] (a staff fragment which contains a certain amount of notes, repeated several times in the same piece) that shall be indicated as themes which might be used for different purposes after having been identified: from stylistic analysis of a composer to the comprehension of musical text in alphanumeric format.

The segmentation process requires the development of tools capable of describing, in mathematical terms, the musical functions most widely exploited by composers of all times within the compositions: tools that nonetheless lead to tackling the problem of "*combinatorial explosion*", related to the potential number of motifs that may be identified by an algorithm. The combinatorial explosion is a real problem within the ambit of the studies on AI and the objective of this article is to represent a mathematical tool able to tackle this problem within the ambit of the analysis of the symbolic musical text, in order to reduce the possible number of identifiable musical objects and, therefore, improve the results of research on them. This analysis tool has been developed taking in consideration Shannon's theory [4], i.e. examining more closely the processes of human choice.

This paper is organized as follows.

Section 2 describes certain theories developed in order to identify motifs within the musical composition. Section 3 describes the methods already used to reduce the combinatorial explosion. Section 4 introduces a new tool meant to reduce the combinatorial explosion. Section 5 shows some experimental tests that illustrate the effectiveness of the proposed method. Finally, conclusions are drawn in Sect. 5.

2   Identifications of the Musical Motif

Musical analysis is not a scientific discipline and, as such, it entails the first difficulty of defining precise objectives of analysis with respect to the various possibilities of exploration. This paragraph illustrates the underlying principles that have so far been used in the various algorithms created for the segmentation of a musical composition.

The segmentation process is made up of three distinct stages. The first stage consists in the acquisition of the staff by the algorithm; the second stage consists in the (actual and effective) segmentation of the musical composition, creating a list of possible musical motifs that will be analyzed in the third stage, i.e. the stage in which, based on specific criteria, the motifs considered important at a musical level will be identified.

A starting point to establish a logical sequence of work and therefore be able to formalize at a mathematical level the analysis tools is represented by a definition proposed by Ian Bent [5] which places the analysis in a "Cartesian Ideal" that, even if unreachable because it is idealistic, provides the background of the motivations that will guide us through the elaboration of the algorithm:

"Musical analysis is the subdivision of a musical structure into relatively simpler constitutive elements and the study of the functions of these elements within this structure".

In musical analysis, one of the important elements is the theme which represents the fundamental motif—often a recurrent one, too—of a composition, especially if it is a far-reaching composition [6]; therefore it is a melodic fragment provided with individuality and recognazibility, often to such an extent as to characterize the entire musical composition [7]. The motifs are never explicitly indicated by the composer on the score, therefore a fragment of notes extracted out of it does not necessarily constitute a motif. This lack of indication grants a mysterious character to the motif itself, as if it was a secret that listening and analysis have the assignment to reveal. In this regard it is necessary to take into consideration a very important aspect of music as far as analysis is concerned, i.e. the psycho-receptive aspect. Reference is made therefore to the *Generative Theory of Tonal Music* proposed in 1983 by Fred Lerdahl and Ray Jakendoff [8] and based on the concept of functional hierarchy. The central hypothesis is that the listener, whose goal is to comprehend and memorize a tonal musical phrase, tries to pin down the important elements of the structure by reducing what he is listening to to a highly hierarchized economical scheme (Fig. 2).

Therefore the idea is that the listener is performing a mental operation of simplification that allows him not only to comprehend the complexity of the surface, but, when it is necessary, also to reconstruct such complexity starting from a simplified scheme and to produce other musical surfaces, other phrases of the same type, through a reactivation of the memorized structure.

The study of these mechanisms lead to the construction of a formal grammar capable to describe the main rules observed by the human mind in order to recognize structures within a musical piece. According to Lerdahl and Jackendoff the grammar they proposed is "language-independent", i.e. such as to function independently of composition styles. Therefore certain rules of this grammar would

Fig. 2 Chorale by J.S. Bach. The example (excerpted from *A generative of tonal music*) reproduces an analysis performed with the Lerdahl-Jachkendoff method. As it can be noticed, the consecutive reductions eliminate an ever larger number of events, reaching the extreme case in which only the tonal chord remains: the information, in this case, is not enough to characterize the phrase, but by limiting himself instead to the first two reductions, a listener would be able to recognize the fragment

qualify as universal of musical perception and could be used to represent innate aspects of musical cognition [9–11]. It is possible to understand many musical works as having been generated by the reverse of this hierarchical reduction process —that is, by the successive elaboration of a fundamental structure with less structural notes, until the detailed foreground or musical surface emerges [12].

So from what has been said so far it can be deduced that in order for the listener to recognize and memorize a certain sequence of notes, it is necessary to present the sequence several times (even if this is one of the motifs that produce the combinatorial explosion [13]).

3 Segmentation and Combinatorial Explosion

Various algorithms for the segmentation of a musical composition were realized on the basis of the theories exposed in the previous paragraph: algorithms structured in such a way as to, first of all, be able to read a score and then be able to analyze it based on specifically provided mathematical operators [14].

Generally speaking the score is seen as a list of numbers each of which corresponds to a sound based on its pitch [12, 15]. If we take into consideration the piano keyboard because it is the instrument with the largest extension, the lowest note, i.e. a natural A, is assigned the value 1 and every single successive sound is assigned an increasing value (A# = 2, B = 3, C = 4, and so on). Thus we obtain the list of sounds as represented in Fig. 3.

A monodic sequence was contemplated in the example in Fig. 4. In case the score is, instead, composed of several voices placed on different staves, the line of reasoning does not change: the list of sounds will be composed first of all by the relative numbers of the first staff, then those of the second staff and so on (Fig. 4) [16].

The score represented this way may be subjected to analysis by an algorithm in order to identify the motifs it contains. The motif might present itself in its fun-

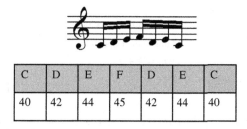

C	D	E	F	D	E	C
40	42	44	45	42	44	40

Fig. 3 List of sounds. The list refers to the above-mentioned fragment. The A column indicates the sound with a Latin notation, while the B column indicates the numeric reference of the sound related to the piano keyboard and the C column indicates the absolute reference

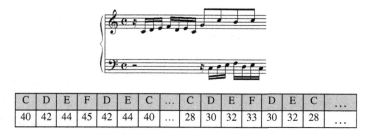

C	D	E	F	D	E	C	...	C	D	E	F	D	E	C	...
40	42	44	45	42	44	40	...	28	30	32	33	30	32	28	...

Fig. 4 Sequential representation of the score

Fig. 5 Invention for two voices in C major BWV 772 by J.S. Bach. The example shows how as soon as in the first three beats, the composer used the theme in its fundamental state (beat 1), the transposed theme (beat 2) and the inverted theme (beat 3)

damental state (that is in the same way it was written the first time), or it could have undergone (see Fig. 5):

- transposition: the process of moving a sequence of notes up or down in pitch by a number of semitones (which represent the transposition degree x) while keeping constant the relative distance between its composing notes;
- inversion: the process through which a melody is transformed by inverting the direction of its single intervals in relation to the initial note, so that every ascending interval becomes descendent and vice versa;
- retrogradation: the process that consists in representing the melody backwards, from the last to the first note.

The necessary tool to describe it in mathematical terms was developed for every one of these cases (see Table 1).

Table 1 Example of the main musical function for melodic analysis

Musical function	Mathematical operator	Musical function
Melodic Transposition	$Tx(h) = h + x$	
Melodic Inversion	$I_i\,(h) = I - (h - I) = 2i - h$	
Melodic Retrogradation	$R(h) = h$ $R(wh) = hR(w)$	
Retrograde of the Inverted	$RI = IR$	

The first column contains the name of the musical function, the second column displays the related mathematical operator and the third column shows an example for every function

The score segmentation process is performed by comparing two sequences of musical notes [16, 17]. In particular, the starting point (I_o) of the segment to search (that shall initially correspond to the first note) is set within the score (represented as a list of number) and its length is established initially the length of the segment shall be 2 (for instance do-re) then 3 (do-re-mi), 4 (do-re-mi-fa), 5 (do-re-mi-fa-re) and so on.

Four interval vectors of analysis representing the segment in its original state (*VectorO*), retrograde (*VectorR*), inverted (*VectorI*) and the retrograde of the inverted (*VectorRI*) are then created and inserted in a matrix the purpose of which is to collect a list of the interval vectors representative of every original segment taken into consideration, in order to avoid performing analysis on a segment the vector of which is equal to the vector of a previously analyzed one (reduction of the combinatorial explosion [17]).

Let us then set the starting point (that shall be indicated by I_s) of the fragment to be compared with the original segment (I_o): starting from point I_s we shall move to the right (one position at a time) for $n - l - I_s$ possibilities (where n is the total number of elements (sounds) inserted in the list and l is the length of the reference segment) considering segments of the same length of the original (Fig. 6) and checking if the interval vector (*VectorC*) of every one of them coincides with one of the four interval vectors (original, retrograde, inverted, retrograde of the inverted) of the original segment.

In case the comparison has a positive result, a representative list of the segment is created and at every turn the starting position of the segment and the type of vector observed (original, retrograde, inverted, retrograde of the inverted) are inserted in the list. This allows us in the end to know, for every segment, how many times (and in which position of the score) it was found in its original, retrograde,

Fig. 6 Example of segmentation

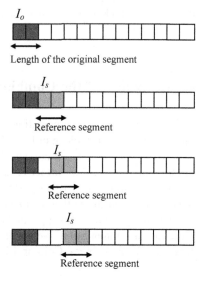

Fig. 7 Example of generation of the original segment

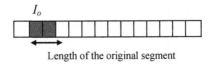

Length of the original segment

Fig. 8 Example of segmentation. Length of the original segment

inverted and retrograde of the inverted state [18]. A general list (*VectorG()*) of the vectors that have a redundancy bigger than one (inasmuch as one is representative of the original vector) is created in the same way: based on the considerations made on repetitiveness (acc. to paragraph III), this segment does not deliver any useful piece of information and is therefore discarded (reduction of the combinatorial explosion [17]).

Once the first original segment was searched within the entire score we move on to perform the analysis, according to the procedure described above, of a new segment that shall be equal to the preceding segment with the addition of one note, then of two notes and so on (Fig. 7).

If the result of the analysis of a fragment is redundancy equals one, then that segment, besides not being inserted in the general list of vectors (see above), determines the closure of the cycle for that segment (the initial position of which is I_o) and the subsequent opening of a new cycle of analysis of a segment the initial position of which is $I_o + 1$ and so on (reduction of the combinatorial explosion [16]) (Fig. 8).

4 Reduction of the Combinatorial Explosion

The objective of the segmentation procedure is to create a list of segments that might represent possible motifs within the score, on the basis of the principles exposed in the previous paragraphs. The number of segments found this way is still too high and in order to further reduce [19] it we can make use of a mathematical tool derived from the set theory: the inclusion. It means checking whether a segment is included in another segment [20].

The procedure is very simple and easy to understand if we consider Table 2, that represents a summary of the segments identified in the analysis of Bach's BWV 772 Invention for two voices. Column 1 indicates the starting position of the segment on

Table 2 Summarizing example of the segments found by the algorithm during analysis oh Bach's BWV 772 Invention for two voices

1	Starting position segment	Segment length	Original segment	Vector
2	1	2	1	O
3	1	2	-1	R
4	1	2	-1	I
5	1	2	1	RI
6	1	3	1 1	O
7	1	3	-1 -1	R
8	1	3	-1 -1	I
9	1	3	1 1	RI
10	...	...	...	...
11	1	5	1 1 1 -2	O
12	1	5	2 -1 -1 -1	R
13	1	5	-1 -1 -1 2	I
14	1	5	-2 1 1 1	RI
15	1	6	1 1 1 -2 1	O
16	1	6	-1 2 -1 -1 -1	R
17	1	6	-1 -1 -1 2 -1	I
18	1	6	1 -2 1 1 1	RI

the score, column 2 indicates its length, column 3 specifies the Original vector (that was compared during the segmentation process) and finally column 4 indicates the type of vector (O = original, R = retrograde, I = inverted, RI = retrograde of the inverted).

For example taking into consideration the segment indicated on row 11 ($V_1 = <1\ 1\ 1\ -2>$) and comparing it to the one on row 15 ($V_2 = <1\ 1\ 1\ -2\ 1>$), we can notice how the two vectors have different lengths, yet by eliminating the last element of the second vector, they become equal.

Data of the two vectors V_1 e V_2:

$V_1 = <1\ 1\ 1\ -2>$
$V_2 = <1\ 1\ 1\ -2\ 1>$

where

$$V_1 \subset V_2 \Leftrightarrow \forall x: x \in V_1 \Rightarrow x \in V_2$$

i.e. V_1 is included in V_2 if and only if, for every element x, if x belongs to V_1 then x belongs to V_2.

In the proposed example we can speak of **strict inclusion** [15], to indicate that every element of V_1 is also an element of V_2, but that there are elements of V_2 which are not elements of V_1.

Once it has been verified that V_1 is included in V_2, it is not possible to cancel the first vector from the list a priori, yet it is necessary to compare the redundancy of both vectors:

- if V_1 and V_2 have the same redundancy, then we must check if the first vector is also found in the same points in which the second is found. If by any chance the two segments are found in different positions then both are valid segments.
- if the redundancy of V_1 is smaller than V_2, we must, nonetheless, check if the first vector is also found in the same points in which the second is found. If by any chance the two segments are found in different positions then both are valid segments.
- if the redundancy of V_1 is bigger than V_2, both segments are valid.

This procedure must be performed between all the segments inserted in the list and its purpose is to reduce the number of vectors that could potentially represent a motif and, therefore, improve the precision of the results of the analysis performed by the algorithm in the stage that comes after the segmentation stage (see paragraph II).

5   The Obtained Results

The model of analysis set forth in this article, based on the principle of inclusion derived from the set theory, was verified by realizing an algorithm the structure of which takes in consideration each and every single aspect described above: the results of the analysis are indicated by means of diagrams that allow an immediate visualization for interpretative purposes.

The algorithm was tested on a small set of musical compositions by different authors and with different styles, written for pianoforte and orchestra.

Table 3 and Fig. 9 display the results of the analysis of certain musical compositions, highlighting for every one of the number of segments identified both without considering the inclusion principle (column 2) described in paragraph IV and considering this principle (column 3). Finally, column 4 indicates the percentage of reduction of the identified segments by applying this principle: a reduction that translates essentially into a major precision in identifying the representative motifs of a musical composition for the melodic analysis algorithms.

Table 3 Results related to the reduction of combinatorial explosion

Musical Work	Number of identified segments (without filter)	Number of identified segments (with filter)	Improvement (%)
J.S.Bach Invention for two voices BWV 772	648	396	39
J.S. Bach Invention for three voices BWV 789	3.608	1.364	63
J.S. Bach Prelude to French Suite in B minor BWV 814	848	284	67
W.A. Mozart First tempo of the Sonata in C major KV 545	8.060	1.732	78
L. van Beethoven Romance for violin and orchestra op. 50	28.708	9.400	66
J. Brahms Intermezzo in C major op. 119	12.112	1.692	86
F. Margola Sonatina n° 2 for pianoforte	3.120	392	87

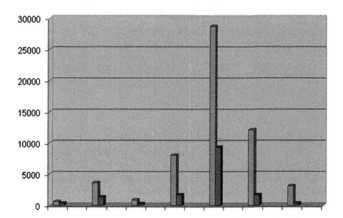

Fig. 9 Graphic representation of the results of Table 3

6 Conclusions and Future Work

In this pilot study, the objective was to reduce the number of motifs that could be identified by an algorithm in the segmentation process of a musical composition, without having to change the analysis parameters on the basis of the style of the specific composition. For this reason we tried to find, first of all, a unique analysis system, valid for any kind of musical forms (from fugues to sonatas, from romances to intermezzos) and for any kind of style (from monody to polyphony); furthermore, we tried to look at the results from a truly scientific perspective, using statistics.

The results obtained are satisfying and do illustrate how it is possible to improve the precision of computer-aided musical analysis, reducing the number of segments to evaluate in the stage following segmentation (Strength): given that the use of information technology in musical analysis does not allow reaching the same results as the ones of an analysis performed by a person. At the same time we should highlight that this procedure increased the time of operation of the computer in the segmentation of the musical composition (Weakness) because it had to perform major comparisons during this specific stage.

The high degree of complexity of musical phenomena imposes certain forms of achievement that must be adequate and that, for completeness' sake, must cope with the problems under a sufficiently large number of angles. Thus, even from a theoretical—musical point of view, the possibility to integrate different approaches appears as a precursory way of interesting developments. And it is really thanks to the new techniques of artificial intelligence that such forms of integration and verification of the results become achievable.

Future work could combine the method presented in this paper with the concept of "tandem repeat type" [21], in order to reduce the amount of musical segments.

References

1. Searle, J.: Minds Brains and Science. Harvard University Press, Cambridge (1984)
2. Russell, S., Norvig, P.: Artificial Intelligence: A Modern Approach. Prentice-Hall International, London (1995)
3. Cordeschi, R.: Artificial Intelligence and Evolutionary Theory: Herbert Simon's Unifying Framework. Cambridge Scholars Publishing (2011)
4. Shannon, C., Weaver, W.: The Mathematical Theory of Communication. University of Illinois Press, Urbana, Illinois (1949)
5. Bent, I.: L'analyse musicale—histoires et méthodes, éditione Main d'Oeuvre, collection musique et mémoire (1997)
6. Della Ventura, M.: Rhythm analysis of the "sonorous continuum" and conjoint evaluation of the musical entropy. In: Proceedings of Latest Advances in Acoustics and Music, Iasi, pp. 16–21 (2012)
7. Fraisse, P.: Psychologie du rythme. Puf, Paris (1974)
8. Lerdhal, F., Jackendoff, R.: A Generative Theory of Tonal Music. The MIT Press (1983)

9. Cambouropoulos, E.: Voice and Stream: Perceptual and Computational Modeling of Voice Separation. Music Perception (2008)
10. Cambouropoulos, E.: Conceptualizing Music: Cognitive Structure, Theory, and Analysis. Music Perception (2003)
11. Cambouropoulos, E.: Voice separation: theoretical, perceptual and computational perspectives. In: Proceedings of the 9th International Conference on Music Perception and Cognition (ICMPC), Bologna, Italy (2006)
12. Velarde, G., Weyde, T., Meredith, D.: An approach to melodic segmentation and classification based on filtering with the Haar-wavelet. J. New Music Res. **42**(4), 325–345 (2013)
13. Lartillot, O.: An efficient algorithm for motivic pattern extraction based on a cognitive modeling. Open Workshop of Musicnetwork, Vienna (2005)
14. Bigo, L., Ghisi, D., Spicher, A., Andreatta, M.: Representation of musical structures and processes in simplicial chord spaces. Comput. Music J. **39**(3), 9–24 (2015)
15. Lonati, F.: Metodi, algoritmi e loro implementazione per la segmentazione automatica di partiture musicali, tesi di Laurea in Scienze dell'informazione. Università degli studi, Milano (1990)
16. Della Ventura, M.: The Fingerprint of the Composer. GDE, Rome (2010)
17. Harald Baayen, R., Hendrix, p, Ramscar, M.: Sidestepping the combinatorial explosion. Lang. Speech **56**(Pt 3), 329–347 (2013)
18. Cambouropoulos, E.: Musical parallelism and melodic segmentation. Music Percept. **23**(3), 211–233 (2006)
19. Grindal, M.: Handling combinatorial explosion in software testing. Ph.D. thesis at University of Skövde (2007)
20. Kolata, G., Hoffman, P.: The New York Times Book of Mathematics: More Than 100 Years of Writing by the Numbers. Sterling, New York (2013)
21. Bose, P., Hermetz, K.E., Conneely, K.N., Rudd, M.K.: Tandem repeats and G-rich sequences are enriched at human CNV breakpoints (2014). doi:10.1371/journal.pone.0101607

Correction to: Solving an Infinite-Horizon Discounted Markov Decision Process by DC Programming and DCA

Vinh Thanh Ho and Hoai An Le Thi

Correction to:
Chapter "Solving an Infinite-Horizon Discounted Markov Decision Process by DC Programming and DCA" in:
T.B. Nguyen et al. (eds.), *Advanced Computational Methods for Knowledge Engineering*,
Advances in Intelligent Systems and Computing 453,
https://doi.org/10.1007/978-3-319-38884-7_4

In the original version of this chapter, a reference to an earlier chapter was omitted. The reference "Piot, B., Geist, M., Pietquin, O.: Difference of convex functions programming for reinforcement learning. In: Ghahramani, Z., Welling, M., Cortes, C., Lawrence, N.D., Weinberger, K.Q. (eds.) Advances in Neural Information Processing Systems 27, pp. 2519–2527. Curran Associates, Inc. (2014)" has now been added.

The updated version of this chapter can be found at
https://doi.org/10.1007/978-3-319-38884-7_4

Author Index

© Springer International Publishing Switzerland 2016
T.B. Nguyen et al. (eds.), *Advanced Computational Methods*
for Knowledge Engineering, Advances in Intelligent Systems
and Computing 453, DOI 10.1007/978-3-319-38884-7

Printed in the United States
by Baker & Taylor Publisher Services